FORWARD

Joan Starr was born and raised in the Tenterfield district of New England. Since retiring in 1983 from full-time journalism and a successful career spanning thirty-six years, she has divided her time between her two grazing properties and her public relations consultancy in Brisbane.

Christopher Sweeney was born and raised in rural South Australia. Shortly after qualifying as a teacher he was conscripted into the army and saw active service in South Vietnam. Following his national service commitment he accepted a commission in the Regular Army. In 1980 he attended the Australian Staff College, Queenscliff. He was subsequently posted as Officer Commanding B Squadron 3rd Cavalry Regiment in Townsville. In 1987 he resigned from the army and currently operates a business on the Gold Coast.

His Royal Highness Prince Charles
Prince of Wales
Colonel in Chief Royal Australian Armoured Corps

FORWARD

THE HISTORY OF THE 2ND/14TH LIGHT HORSE
(QUEENSLAND MOUNTED INFANTRY)

JOAN STARR

CHRISTOPHER SWEENEY

University of Queensland Press

First published 1989 by University of Queensland Press
Box 42, St Lucia, Queensland 4067 Australia
Paperback edition published 1990

© Joan Starr and Christopher Sweeney 1989, 1990

This book is copyright. Apart from any fair dealing
for the purposes of private study, research, criticism
or review, as permitted under the Copyright Act, no
part may be reproduced by any process without written
permission. Enquiries should be made to the publisher.

Typeset by University of Queensland Press
Printed in Australia by The Book Printer, Melbourne

Distributed in the USA and Canada by
International Specialized Book Services, Inc.,
5602 N.E. Hassalo Street, Portland, Oregon 97213-3640

Cataloguing in Publication Data

National Library of Australia

Starr, Joan, 1927- .
  Forward : the history of the 2nd/14th Light Horse
  (Queensland Mounted Infantry).

  Bibliography.
  Includes index.

  1. Australia. Army. Light Horse Regiment, 14th. 2.
  Australia. Army. Light Horse Regiment, 2nd. 3. World
  War, 1914–1918 – Regimental histories – Australia. I.
  Sweeney, Christopher, 1946- . II. Title.

357'.0994

ISBN 0 7022 2304 2

*This book is dedicated to all
who served in the Regiment.*

Metric Conversion Table

Readers wishing to convert Imperial measurements into present day metric equivalents should use the following conversion factors:

1 mile	=	1.6093 kilometres
1 yard	=	0.9144 metres
1 foot	=	0.3048 metres
1 inch	=	25.3999 millimetres
1 ton	=	1.016 tonne
1 stone	=	6.35 kilograms
1 pound weight	=	0.4536 kilograms
1 ounce	=	28.3495 grams
1 gallon	=	4.546 litres

Currency is expressed in the Imperial Australian and Sterling units of the time, pounds, shillings and pence (£.s.d).

Contents

List of Illustrations *ix*
List of Figures *xi*
Foreword *xiii*
Acknowledgments *xv*
Glossary *xvii*

1 Queensland Colonial Soldiers *1*
2 Soldiers of the Queen *11*
3 The Great War *55*
4 Sand, Sweat and Horses *81*
5 Their Finest Hour *107*
6 Training Hard *163*
7 Out of the Saddle *179*

Appendixes
1 Honour Roll of the Queensland Mounted Infantry *195*
2 Honour Roll of the 2nd Light Horse Regiment *197*
3 Honour Roll of the 14th Light Horse Regiment *204*
4 Battle Honours *205*
5 Awards for Gallantry *207*
6 Order of Battle *212*
7 Commanding Officers *213*

Notes *215*
References *219*
Index *223*

Illustrations

HRH Prince of Wales *frontispiece*

following page 62

Shearers' Strike 1891
Range practice 1890s
Queensland Mounted Infantrymen
QMI banner presented by the Lady Mayoress 1899
1st Contingent departing for the Boer War
Horse stalls aboard the ship
Memorial at Sunnyside 1900
Replica of the Sunnyside memorial erected at Enoggera 1989
Private V.S. Jones, killed at Sunnyside
Lieutenant Colonel Pilcher and officers
Soldiers with the Union Jack at Elands River
A stone recording the names of four men killed at Elands River
Memorial at Anzac Square, Brisbane
Recruits at Chermside 1914
John Markwell farewelling his son "Willie"
2nd Light Horse Regiment embarked for Egypt
Light Horse camp 1915
2nd Light Horse Regiment officers
Cameliers and their camels
Ready for action
2nd Light Horse Regiment in the Middle East
Field service postcard 1916
Greeting card from Sinai 1916

Horses grazing near Gaza
Ghoraniye defences
Dummy horses in the Jordan Valley
Creating a dust storm

following page 150

Musallabeh, or "Camel's Hump"
The 2nd Light Horse Regiment at Musallabeh
Turkish soldiers near Jaffa
Valley of death
Scene of destruction in the Barada Gorge
Presentation of the 2nd Light Horse Regiment Guidon
Presentation of the 14th Light Horse Regiment Guidon
Harrisville Troop with the Prince of Wales Cup
The Boonah Troop
The Toogoolawah Troop
The Laidley Troop
The MMG Troop
Prince of Wales Cup competition
The 1st Cavalry Brigade on parade March 1940
The Regiment's final mounted parade at Southport 1940
On the beach
Machine-gun carriers
Watering lines at Black River 1942
Yorkforce in North Queensland 1942
Coral Sea parade in Brisbane 1950
Prime Minister's escort in London 1953
Prince of Wales Cup winners 1953
106mm RCL firing at Tin Can Bay
M113A1 being unloaded at Bulimba
Sharing a joke, camp 1973
Shoalwater Bay 1985
An original lighthorseman on Anzac Day

Figures

1. Queensland   4
2. South Africa 1899   12
3. South Africa, the area affected by the Boer War   22
4. Battle of Elands River   40
5. Order of Battle 1914–1918   58
6. Gallipoli   64
7. Middle East   77
8. Defensive battle at Romani August 1916   87
9. Battle of Beersheba   115
10. Jordan Valley   122
11. Ghoraniye Bridgehead   126
12. Battle of Abu Tellul   135

Foreword

It is a pleasure to write a few words of introduction to the fascinating story of the gallant band of Australians who formed the Queensland Mounted Infantry. In 1904, the then Prince of Wales presented a trophy for competition between the outstanding Light Horse Troops from each state of Australia. For more than fifty years the Cup was competed for, ensuring that the soldiers of the Light Horse maintained the high standards achieved on the battlefields of South Africa, Gallipoli, Sinai and Palestine.

During the Boer War the Queensland Mounted Infantry established their reputation on the battlefields of Sunnyside, Kimberley, Paardeberg, Diamond Hill and Elands River. At Sunnyside on 1 January 1900, the Queensland Mounted Infantry achieved the sad honour of losing the first two Australian soldiers killed in action. British historian Sir Arthur Conan Doyle wrote of the defence at Elands River, "there was no finer fighting in the war".

The reputation gained in South Africa was maintained by the 2nd and 14th Light Horse Regiments during World War I. On Gallipoli the 2nd Light Horse was assigned to Quinn's Post, a most difficult sector of the Anzac line. The Regiments played a prominent part in all the successful operations in Sinai and Palestine. At Romani it was undoubtedly the stubborn resistance of the Regiments of the 1st Light Horse Brigade (the 1st, 2nd and 3rd Light Horse) against the Turkish attack that won the battle, thereby changing the campaign from one of a defensive nature to the offensive. In the Jordan Valley, early in the summer of 1918, the 2nd Light Horse bore the brunt of two German and Turkish attacks at Ghoraniye and Abu Tellul. In the latter battle the gallant resistance of Light Horse posts, surrounded on all sides by the enemy, saved the situation. The resistance by the 2nd Light Horse on this occasion is regarded by many as the finest fighting performed by the Light Horse.

The Regiment today is still in the forefront, being the first of the Australian Army to become a fully integrated unit of both Regular Army and Army Reserve soldiers. I commend this history to the people of Australia and particularly those of Queensland, of whose horsemen the members of the Regiment were so typical.

W. B. Campbell

Sir Walter Campbell
Governor of Queensland

Acknowledgments

Many serving and ex-members of the 2/14 Light Horse Regiment, and people closely associated with it, have helped us to gather material for this book and to check and re-check information. To name but a few: the late Major Syd Appleby, whose tales brought vividly to life the lighthorsemen of the past, and the deeds which gave them an unrivalled place in history; Captain Wayne Hickey, who handed over a great deal of material he had painstakingly researched; Lieutenant Phil Rutherford, who travelled many miles to gather records and pictures from ex-troopers or their families; that enthusiastic hoarder of historic information, retired Major Len Friend; Trevor Langtry, a past Regimental Sergeant Major, who gave constant encouragement and assistance; Lieutenant Colonel Tom Childs, last Commanding Officer of the Regiment before integration; Lieutenant Colonel Paul Coleman, first Commanding Officer of the Regiment after integration; Lieutenant Colonel Chris Stephens, Commanding Officer of the Regiment at the time of publication; Major Graham Tregenza, Captain Graham Jardine-Vidgen, Captain Bob Burn, and Brigadier Colin Wilson, all of whom have supplied needed information and have helped check our interpretation of it; Colonel Bill Morgan, Lieutenant Colonel Miles Farmer, Captain Ron Squires, and Major Kim Ninnes; Noel Basset and his wife Val who took over the task of cataloguing photographs and of copying them for us; Ken Kindness, who went to work at the War Memorial in Canberra during the writing of the book, and forwarded much valuable information; Monica Kortz, one of our most enthusiastic researchers; Colonel Des Kelly, first President of the Book Committee which started the project off and has worked hard to make it a success; Captains Darryl Proud, Tony de la Fosse, Paul Williams, Ken Corke, and all the other officers who prepared information about the battle honours borne by the Guidons of the 2nd and the 14th Regiments; the Survey Section Headquarters 1st Division and 1st Field Survey Squadron for the preparation of maps and illustrations; Roy Warren, Col Head and the others of the Boonah and Kalbar Troops, with whom we spent many informative and enjoyable hours learning of the peacetime activities of the Light Horse groups.

There are a number of sponsors whose assistance ensured the story of this Queensland Light Horse unit would be told. Firstly, special thanks to Castlemaine Perkins for their contribution towards advertising and publication. Sponsors include the Queensland Branch RAAC Association, Boonah Light Horse Association, Longreach RSL, Cooktown RSL, Toowong RSL, Yeronga Services Club, Geebung–Zillmere Services Club, Carlton and United Breweries, QUF Industries, GWA Ltd, Rockhampton TV, G.J. Alford, Mathers Shoes Ltd, N.H. Plunt (Bank of Queensland), Campbell Brothers, Pioneer Sugar, PNQ Investments, Mackay TV, G.W. Ltd, F.J. and C.L. Morgan, C.J.A. and N.M. Sutton, G. Jardine-Vidgen, Officers and Sergeants Messes 2/14 Light Horse (QMI), Beersheba Club, and many small donors. Thanks also to Retusa Pty Ltd for permission to use an extract from the works of Banjo Paterson.

And most of all we would like to thank Major Chad Sutton, without whose help this book would probably never have been completed. Many indeed were the hours Chad spent researching, marking and identifying passages for us in books, letters, diaries, and the mass of printed information which he and others found from many sources. He was always available to discuss any aspect of the unit's history. Our sincere thanks to him. We hope he and all the others enjoy the result of a real team effort.

Assistance and support has come from many sources. If we have not mentioned all the people who helped us in some way, we hope we will be forgiven.

Our own part in putting all the information together is our gift to the Regiment. It is given with deep affection and respect.

Glossary

Boer	farmer; during the Boer War implied anyone who fought against British and Allied troops
burg	town
burgher	a man of one of the Boer Republics with full citizen rights
cacolet	wicker chair strapped to a camel for use as an ambulance
CB	Companion of the Order of the Bath
CMG	Companion of the Order of St Michael and St George
commandant	senior officer of a Boer commando
commando	irregular unit of mounted Boers
DSO	Distinguished Service Order
drift	ford
FOO	forward observation officer; an artillery officer who moves with the forward troops in order to direct the artillery fire support
KCMG	Knight Commander of the Order of St Michael and St George
kopje	small hill
laager	Boer camp with vehicles usually lashed together in a circle to form a protective barricade
MC	Military Cross
MID	Mentioned in despatches
MM	Military Medal
Mauser	German rifle used by the Boers
mealie	maize on the cob
nek	saddle or junction point joining two hills
pom-pom	one-pounder automatic Maxim gun with twenty-five shells on a belt
picquet	small body of soldiers stationed out from a force to warn against an enemy's approach
RMO	Regimental Medical Officer
rand	ridge; "The Rand" is the area around Johannesburg
SAA	small-arms ammunition
sangar	low stone breastwork or rifle-pit for several men
Uitlander	foreigner; British settler in the Boer Republics
veldt	open country bearing grass, bushes or shrubs or thinly forested
wadi	rocky watercourse dry except in the rainy season

Go, live as men live, or to die as men die!
In the front of the battle be heard your war cry!
Let the grey emu plumes wave aloft with renown!
Or be found on the field where the bravest went down!

A. Meston [1]

Queensland Colonial Soldiers

Origins

From the landing of the First Fleet to the late nineteenth century, Australia was a temporary home for over a quarter of the line regiments of the British Army. In the days when the Regiments of Foot were still numbered, twenty-six of them served in the Australian Colonies. Apart from ceremonial duties their responsibilities were largely of policing: guarding convict chain-gangs and jails; protecting settlers from the Aborigines and bushrangers; and protecting the goldfields and gold consignments, and so on. Although many Australian born men were recruited into the British units stationed in Australia, the colonies' defence and internal law were in the hands of the British Army.

In the mid-nineteenth century, as British units were disbanded and the troops withdrew from Australia, the colonies faced the task of raising and funding a standing army of their own, and of coordinating a defence policy. Following the separation of the Colony of Queensland from New South Wales in June 1859, steps were taken to provide for its defence. The responsibilities assumed by the new and sparsely populated colony included the defence of its 3,000 mile coastline, and its vast 670,500 square miles – an area as large as Great Britain, Ireland, France, Spain, Portugal, and Italy combined. For a population of only twenty-five thousand, almost one third of which was concentrated in the southeast corner in Brisbane and Ipswich, this was a formidable task. Furthermore the small detachment of Imperial troops had been withdrawn some years earlier so the Colony lacked even the nucleus of a defence organisation upon which to build.

In early February 1860, the Colonial Secretary announced in the *Moreton Bay Courier* that lists were open for the enlistment of one troop of twenty-five Mounted Rifles and two companies of fifty Riflemen each. All men were to be volunteers and the Mounted Rifles were to supply their own horses and uniforms, the government undertook the supply of weapons and ammunition. The volunteers were to be paid only if called out for service and provision would be made for the families of any men killed in action. It was considered that, "two hours daily on two days of the week would suffice for drill and practice, or even Sunday evenings alone. Such practice would be a healthy and agreeable exercise and would promote sociability and good feelings."[2]

The birth of the unit that is now known as the 2nd/14th Light Horse (Queensland Mounted Infantry) occurred on 27 February 1860, when the Governor, Sir George Bowen, approved the Rules and Regulations of the Brisbane Mounted Rifles. The Governor appointed John Bramston as Captain and commander of the troop. Later that year two more troops of the Queensland Mounted Rifles were raised and known by their district titles of Ipswich and Port Curtis.

Following an enthusiastic start the Volunteer Defence Force experienced difficulty retaining its recruits. Within two years the numbers in certain units dwindled due to a lack of interest, and dissatisfaction caused by the government's failure to provide sufficient uniforms and equipment. In March 1863 the Governor dispensed with the services of the Port Curtis troop due to their failure to attend parade, the commissions of the officers were cancelled. The Queensland Mounted Rifles were renamed Queensland Light Horse in 1864, but their numbers continued to dwindle with the troops at Brisbane and Ipswich numbering twenty-five. The *Moreton Bay Courier* noted that they were the most irregular troops of an unorthodox army. The men were said to be loath to accept discipline and their only usefulness was for the purpose of show. The paper predicted an early disbandment of the remaining mounted troops. Unfortunately, their prediction was accurate and by 1866 both the Brisbane and Ipswich troops had been disbanded. Queensland had chosen, for reasons of economy, to create an amateur defence force. It would take many years for the government to realise its mistake, but in the meantime the Queensland Defence Force continued under difficulties and by 1876 the total number was only 415 men enrolled in artillery and infantry units.

This was all to change as fears of a Russian invasion and concerns over unrest in the Torres Strait region caused a government rethink of its defence policy. In September 1883 an Imperial officer, Lieutenant Colonel George Arthur French, Royal Artillery, was appointed as Commandant of Queensland's Defence Force. French was a committed militiaman who wanted to eliminate the volunteer force completely and create a militia force, or at least one that was partially paid. Although he was not allowed a completely free hand in reorganising the force, French did however generate considerable change in the government's attitude to defence. A professional, he expressed himself forcefully and he would not be ignored when presenting his reports criticising a lack of government action. He had no intention of tolerating indifference and his dynamism had effect.

The Queensland Defence Force was substantially reorganised under the 1884 Defence Act with £35,591 allocated for the 1885-86 year. Under the reorganisation the new Volunteer Corps was gazetted on 4 March 1885. The old volunteer units, with the exception of a few in the country areas, were converted to militia, or paid army. The militia component of the Defence Force had a strength of 110 officers and 1,528 other ranks.

French distinguished between the paid and unpaid forces by calling the militia "Defence Forces" and labelling the unpaid troops "Volunteers". Within three years he had increased the Volunteer Corps to 26 officers and 718 other ranks. Included in the volunteers were the Brisbane Mounted Infantry (renamed Moreton Mounted Infantry in 1885), Bundaberg Mounted Rifles, Gympie Mounted Rifles, Mackay Mounted Rifles, and Charters Towers Mounted Infantry. The 1884 Defence Act and Lieutenant Colonel French had provided Queensland with a stable defence structure, among its finest troops were the mounted soldiers.

The Shearers' Strike

The first "action" for the Queensland Defence Force occurred in 1891 in the Western District of the Colony in what was to become known as the "Shearers' Strike". The confrontation began in the late 1880s when leading pastoralists, feeling the pinch of lower wool prices announced they would reduce the pay of the shearers. The shearers were already seething at their unjust conditions, they were paid as low as eight shillings per hundred sheep and then had to pay unusually high prices for the goods provided by the squatters, often leaving the shearers as little as

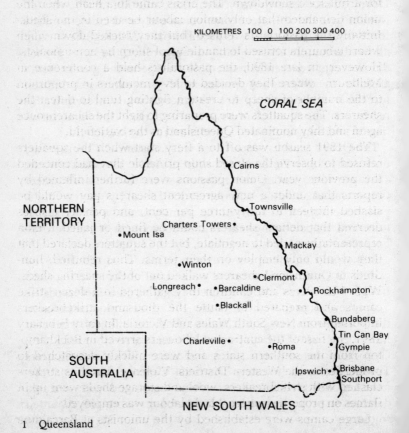

1   Queensland

fifteen shillings per week. For this they worked eleven hours per day and were accommodated in the roughest of conditions — earth-floored huts without ventilation or lights. They could not quit for fear of forfeiting accumulated pay, but the squatter on the other hand could withhold payment if he considered a sheep not properly shorn, and he could dismiss a worker without reason and not pay a penny for work already completed. To counter these unfair conditions, the shearers formed unions and between 1886-89 staged more than three hundred minor strikes and walk outs.

In 1890, the Queensland Shearers' Union was strong enough for a full-scale showdown. The crisis came to a head when the union demanded that only union labour be used in the sheds. Initially the pastoralists refused, but they backed down when wharf labourers refused to handle wool shorn by non-unionists. However, in late 1890, the pastoralists held a conference in Melbourne where they decided to levy members in proportion to the number of sheep to create a fighting fund to defeat the shearers. The squatters were preparing to fight the shearers once again and they nominated Queensland as the battlefield.

The 1891 season was off to a fiery start when the squatters refused to observe the closed shop principle they had conceded the previous year. Union passions were further inflamed by reports that under a new agreement shearer's pay would be slashed thirteen to thirty-three per cent, and penalty clauses decreed that defiant shearers could be fined or jailed. Union representatives tried to negotiate, but the squatters declared that they would only employ on their terms. Thus rebuffed, hundreds of Queensland shearers walked out of the shearing sheds. With their wives and children they gathered in a dozen strike camps and prepared to battle the thousand strikebreakers imported from New South Wales and Victoria. In early February 1891, the first of the contracted labourers arrived in Rockhampton from the southern states and were quickly despatched to properties in the Western Districts. Violence flared as strikers clashed with strikebreakers, wool and storage sheds went up in flames on properties where "black" labour was employed.

Large camps were established by the unionists at Barcaldine

(1,000 men), two near Clermont (350 and 150 men), and Capella (80 men). Inflammatory speeches were made with threats of kidnapping squatters and their families, and burning properties. At this juncture the Queensland government ordered additional police to the troubled areas and police magistrates were instructed to enrol all available men as special constables. There was a fear that the trouble could spread and erupt into civil war.

By 20 February the situation had reached such a serious stage that the government decided to call out the Defence Force in aid of the civil power. That morning the Officer Commanding the Defence Force, Major Jackson, was warned to prepare for immediate embarkation. At 5.00p.m., a force of sixty-one men complete with weapons, a machine-gun and a 9 pounder field gun, sailed for Rockhampton. Three hours later orders were issued to call out the Moreton Mounted Infantry. The Commanding Officer, Major Percy Ricardo, took immediate steps to secure the attendance of all members of the units resulting in four officers and fifty-five men being selected for active duty.

They embarked on the SS *Wondonga* and sailed for Rockhampton the following day. No horses were despatched as the Pastoralists' Association had arranged to provide mounts, but every member took his own saddle and bridle in addition to kit, accoutrements and ammunition. That same day, 21 February, the Rockhampton Mounted Infantry was called out by the local Police Magistrate. On arrival at Rockhampton the entire force came under the command of Major Jackson. The force travelled by train to Clermont and thence despatched to farms to protect the free labourers and property.

As the situation continued to deteriorate the government called out additional volunteer units including the Wide Bay Mounted Infantry, Mackay Mounted Infantry, Darling Downs Mounted Infantry, Charters Towers Mounted Infantry, and Townsville Mounted Infantry, in addition to units of the Defence Force. The task of aid to the civil power is one that most soldiers dislike. It was work that neither commanders nor men had been trained for and which demanded a level of discipline to be found only among the regulars. Nevertheless the men performed the duty admirably as evidenced by Lieutenant Harry Chauvel's action at Charleville.

Chauvel received a warning order on 24 March to be ready to move to Charleville with twenty men and their horses. The men of the Darling Downs Mounted Infantry were issued with fifty rounds of ammunition each, and were on their way by rail the following morning. By this time there were reports of the burning of pasture and fences by the unionists, while at Blackall a non-unionist's bullocks had been shot.

Chauvel was sent to escort a party of free labourers through the bush to a property north of Charleville. A few police were added to his command. It was a miserable journey across black soil in pouring rain so that after twenty miles the men and horses were exhausted. When barely a mile from their destination, the party ran into a crowd of about two hundred shearers streaming down the track. Several of them were wanted by the police and the inspector in charge quickly arrested four of them.

The shearers closed around the party, some waving iron bars and clubs, shouting in their excitement. One of the four men arrested began inciting the shearers and the situation became dangerous. Chauvel gave the order to load; the inspector told him to force his way through the crowd with his troops and free labourers. The raised rifles of the Mounted Infantry had a cooling effect on the tempers as they shepherded the non-unionists and four prisoners along the road. The inspector thwarted an attempt by the angry shearers to follow up and Chauvel reached Oakwood where another detachment of Mounted Infantry was waiting; these set off after the shearers and more arrests were made. Although minor, the incident impressed on Chauvel the power of discipline and the importance of a cool head in a crisis. Had he not kept his Mounted Infantry under strict control, and they were after all mainly station hands with little military training, there may have been a tragedy.

The Mounted Infantry units were employed principally on patrol and escort duty and to keep communications open between detachments, the police and headquarters. Strong detachments were posted at stations where free labourers were shearing and small detachments were posted as sentries at woolsheds and station buildings. The infantry and artillery were responsible for the security of public buildings, jails, railway stations, and goods sheds.

The tension reached its peak when 200 troops swooped on the strike committee's headquarters at Barcaldine and arrested twelve of the leaders, charging them with conspiracy. The strikers were outraged, some men calling for revolution. At Gympie soldiers fixed bayonets to disperse a menacing crowd, while at Rockhampton 200 strikers heckled police guarding the twelve arrested at Barcaldine, when they came to trial. During the trial, the judge Mr Justice Harding was scarcely impartial, stating that he would have shot the strikers if he had been one of the police. He sentenced the twelve including George Taylor and William Hamilton, who later became members of the Queensland Parliament, to three years hard labour each. These severe sentences provoked another outburst of violence.

Although the strikers voted to stay out on strike, signs of weakness began to appear. The first crack came when threats of long-term sanctions by the squatters forced wool carriers back to work. They really had no alternative, for with whole families subsisting on ten shillings a week, men, women and children were on the verge of starvation. Furthermore, rain had turned the strikers' camps into quagmires of stinking mud. On 11 June 1891, union leaders announced that the strike fund was exhausted. The strike was over. Although the shearers had lost their fight, many claimed that in the long term it had led to victory, as the bitter defeat convinced the unions of a need for a political Labor Party to fight their cause in Parliament.

The strike had proven to be an expensive affair, costing the government £170,000, the pastoralists £41,000 and the unions £60,000, huge sums in 1891.

At times the work of the Defence Forces had been arduous, each district being in a state of flood for a considerable period. Colonel French noted:

> Long and trying marches, by day and night, over boggy country (swimming rivers several times a day), were constantly made by the Mounted Infantry, notably so by a detachment of the Moreton Mounted Infantry, under Lieutenant R.S. Browne, which marched 109 miles in thirty-two hours, and by a detachment of Darling Downs Mounted Infantry, under Captain King, which marched sixty-five miles in one day on grass-fed horses.[3]

Colonel Drury reported, "The general conduct of the troops called out has been reported by all Commanding Officers to have been good. . . . To the discretion and judgment shown by officers in command, and the patience exhibited by all ranks under provocation and insult, must be credited the fact that bloodshed, or injury to life or limb, has been happily obviated".[4]

As the strike wore on, the men were out on patrol for about five months, and for most of their service endured the boredom that is so often the lot of soldiers. On the plains of Western Queensland emu were plentiful and the men could not resist the excitement of riding after a quarry that could give them a chase at speed. Emu feathers, tucked into the men's felt hats began to appear and soon became widespread among the soldiers. Bill Lieshman of the Gympie Mounted Rifles claimed to be with the group who started the practice. Writing many years later Bill said:

> I was in a patrol under Lieutenant Vivian Tozer of the Gympie Mounted Infantry, at Coreena Woolshed. On the way we met another Gympie Mounted Infantry patrol under Captain W. Shanahan and they were chasing an emu, which came toward us. When it was shot, some of us dismounted and Terry Rogers and myself were the first to pull the tail feathers out and place them in our hats. Then all in the patrols got the feathers and placed them in their hats.[5]

When they returned home the Queensland Government allowed the Mounted Infantry to wear the emu plume in recognition of its service during the strike. At first it was solely a Queensland decoration, but in 1903 the privilege was extended to Tasmanian and South Australian regiments and finally, in 1915, to all regiments of the Light Horse.

There's a very well-built fellow, with a swinging sort of stride,
About as handy sort as I have seen
A rough and tumble fellow that is born to fight and ride
And he's over here a-fighting for the Queen.

He's Queensland Mounted Infantry compounded 'orse and foot.
He'll climb a cliff or gallop down a flat.
He's a cavalry to travel but he's infantry to shoot.
And you'll know him by the feather in his hat!

Banjo Paterson [1]

Soldiers of the Queen

The Background

In the mid 1600s the Dutch East India Company established trading posts on the Cape of Good Hope to provide a port enroute to the Indies. The original settlers had fled religious persecution in their own countries, of Holland, Germany and France. They were all believers in the fundamentalism of the Bible and sought another Eden in South Africa. They discovered a rich land of fertile soil, abounding in wild life; and there were black men to serve them. The Cape colony prospered as the farmers – the local word was Boer – claimed more land and slowly spread inland.

Affairs in Europe were to have a devastating effect on the colony! During the Napoleonic Wars, Holland allied herself with France. To prevent the sea route to India falling under French control, Britain occupied the Cape for eight years. When hostilities ceased in Europe the British handed the territory to the officials of the Dutch government, so ending the rule of the Dutch East India Company in South Africa.

When war once again flared in Europe, Britain, with the indifference of the world's dominant power, once again annexed the Cape. However, when peace came in 1814 she retained the Cape because of its strategic importance on the sea route in India. The Dutch government was compensated by the payment of £6 million. The people who now came under British rule were no longer Dutch, German or French but a distant race of Afrikaaner people with strong loyalties of their own. At the time of transfer to Britain the white population numbered twenty-six thousand.

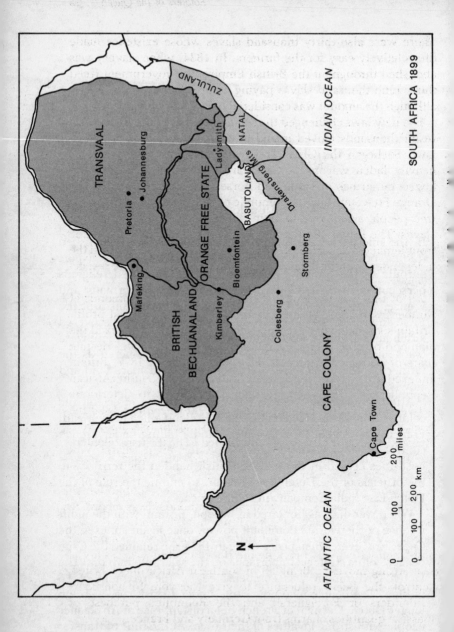

There were also thirty thousand slaves whose existence made life relatively easy for the farmers. In 1834, when slavery was abolished throughout the British Empire, the government freed thirty-nine thousand slaves paying compensation to the Boers, although the amount was considered below market value.

The new laws challenged the social tradition of the Boers. As a result thousands moved inland to escape the control of British Law. So began the Great Trek which lasted sixteen years as heavily laden wagons – Cape-carts – lumbered inland, followed by strings of cattle and horses. The Boers founded the Orange Free State and the Republic of the Transvaal. They were free people again, free to farm, to hunt, to pray, and to own slaves. The British did not oppose these migrations because the troublemakers departed. In 1852 and 1854 Britain recognised the independence of the Transvaal and the Orange Free State respectively.

All of this was to change with the discovery of diamonds at Kimberley in 1867. Because there was likely to be great profit, Britain soon made her presence felt. In 1872 she re-annexed the diamond-bearing region from the Orange Free State, despite roars of outrage and protest. Five years later, in 1877, Britain annexed the recognised and independent South African Republic, the Transvaal. Relations continued to deteriorate between the British and the Boers. On 30 December 1880 the Boer Republic was proclaimed. Several small British units were attacked and annihilated or imprisoned. The British relented, granting independence to the Republic of South Africa – under British sovereign power.

The Boers had an internationally recognised republic which with its vast supplies of gold and diamonds, was about to become the world's richest state. Tension increased between the Boers and the British due to: Britain's annexation of Zululand and Natal, effectively blocking the Boer Republic's access to the sea; strong moves to bring all of southern Africa under British control; the Boers' reluctance to give the vote to non-Boers (Uitlanders or Foreigners); and the Republic's purchase of massive quantities of arms from Germany and France.

War

The situation was inflamed when the British South Africa Company mobilised its private army of eight hundred men and held them on standby at Mafeking ready to ride to Johannesburg in the Republic of South Africa. On New Year's Day 1896, a shocked world learned that this force was riding into Transvaal to take armed support to the Uitlanders who, it was claimed, had asked for assistance. It was a weak display as the force of eight hundred surrendered to the first Boer troops which confronted them.

The British continued to threaten the Republic by moving troops from the Cape to the Transvaal border and ordering reinforcements from India. On 2 October 1899 President Kruger of the Boer Republic issued an ultimatum to the British, accusing Britain of interfering in the internal affairs of Transvaal and of massing troops to threaten the state. The statement concluded, "That unless Her Majesty's Government complies within 48 hours the Government of the South African Republic (Transvaal) would with great regret be compelled to regard the action as a formal declaration of war."[2]

War began on 14 October 1899 when the Boers swept out of the Transvaal and the Orange Free State on three fronts in an effort to deliver the most telling blows before the arrival of more British troops. In the west they besieged the towns of Mafeking and Kimberley, at the same time severing the road to Rhodesia. In the east they seized the northern Natal towns of Dundee and Newcastle and encircled Ladysmith. The third thrust took place in the centre. There they occupied the railway town of Colesburg, severing the rail system from the Cape Colony. The early battles took place on British territory and it was to be several months before fighting occurred in the republics. The war was only a month old and the Boers had drawn first blood and had laid seige to Mafeking in the north, Kimberley to the west and Ladysmith to the east. Britain was outgunned, outmanned and outclassed!

Mobilisation

It is not possible to determine the exact number of men in the irregular Boer forces. Approximately 50,000 men from the two Boer republics, augmented by 10,000 rebels from the Cape Colony and Natal and about 2,500 foreign sympathisers were pitted against the British. They were not all in the field at the one time, the greatest number being about forty-five thousand in December 1899 after which the number gradually declined. Following the fall of Pretoria in June 1900 the number on active service never rose above twenty thousand. After the first ten months of the war, the Boers fought in a purely guerilla fashion. In the end it was to take 448,000 British and Empire troops to subdue the Boers.

Britain was generally out of favour with the rest of the world because of its bullying tactics against the independent states of South Africa. There had been unfavourable comment both in America and much of Europe (Germany in particular), about Britain's actions in South Africa, but it still had the Empire's colonies and dominions on which to call. Even as early as July 1899, when it first appeared that hostilities would break out, an overeager government of Queensland, acting on a recommendation of the Commandant of Queensland Defence Forces, offered a contingent of mounted infantry with a machine-gun section. The offer was matched at once by the governments of New South Wales and Victoria.

The British War Office had no understanding of the value of these so readily offered colonial troops. The soldiers were assessed as less than first-class and certainly not as skilled or as reliable as British regulars. While the War Office was inclined to send a carefully worded polite refusal, the British government's demand for a show of Empire unity took precedence. Under Cabinet pressure, the War Office accepted the offer from the Australian colonies.

Patriotic Response

The Queensland Legislative Assembly debated the whole matter of the offer of troops for South Africa, during a four-day sitting from 11 October 1899. Finally it was decided that the contingent would comprise 250 mounted infantry and one machine-gun section and that the cost to raise, equip and transport the force for a six-month expedition, would be £32,000. On 12 October, during the Legislative Assembly debate, it was learnt that war had been declared by the Boers. On the following day the Boers crossed the Natal border. The British Empire was at war. The Queensland contingent was enrolled, organised and equipped with arms, clothing, horses, saddlery, transport wagons with fittings, and stores between October 13 and October 28 – a praiseworthy record of the indefatigable way in which the staff, the Comptroller of Stores, the medical and veterinary authorities, and the government, strove to equip the contingent for war.

The Queensland Government also paid for another two contingents, the 2nd Queensland Mounted Infantry commanded by Lieutenant Colonel K. Hutchison, and the 3rd Queensland Mounted Infantry commanded by Major W. H. Tunbridge. In addition the Imperial Government paid the expenses of three contingents, the 4th, 5th and 6th Queensland Imperial Bushmen, largely consisting of volunteers from the Queensland Mounted Infantry and other units of the Queensland Defence Force. Following Federation the Commonwealth Government despatched eight battalions of the Australian Commonwealth Horse Regiment. The 7th Battalion of this regiment was raised entirely in Queensland in 1902, and was commanded by Lieutenant Colonel Harry Chauvel. Queensland also provided a company for the 1st and 3rd Battalions of the Commonwealth Horse.

The government decided it was necessary to insure the lives of the men who were about to go to war, and on November 1899, the *Queenslander* reported "The Government has definitely decided to accept the offer of the Mutual Life Association of Australasia to insure the lives of the members of the Queensland contingent for £250 each. The risk is to commence from the date of arrival in South Africa and to continue for a period of 12 months, or until the termination of the war or the departure of the contingent from South Africa."[3]

Privately subscribed funds were raised by citizens throughout Australia, who felt that by authorising despatch of the contingents, the government and people of this country had incurred a moral obligation to the men, their wives and families. It was considered in the event of death or permanent disability of any man in the contingents, his wife and children should be entitled to receive an allowance from what was known as the National Patriotic Fund. To establish the fund, meetings were held in the various capital cities, and as enthusiasm for the cause raced across the country almost every town and settlement took part in fund-raising. The Patriotic Fund was to continue in existence until the 1980s, helping families of servicemen who had suffered as a result of war or natural disaster.

As well as arranging insurance for the men, the government made provision for families during the absence of the breadwinners, by arranging for married men to assign their pay in favour of their wives before leaving for South Africa. Because of these provisions, the men were able to leave their families with the knowledge that they would be well-cared for, should the need arise, by the government and their fellow Australians.

Off to War

Departure of the troops on the *Cornwall* in November drew vast crowds to the wharf area, and the whole event blossomed into a gala occasion with the bands, bunting, garlands, refreshment booths, and bright sunshades. Almost every craft capable of floating, large and small, and gaily decked-out with flags and bunting, gathered not far from the *Cornwall*. The contingent commander, Major Ricardo, led his men onto the wharf where he dismounted. First to march on the wharf was the machine-gun section (a detachment of the Queensland Royal Australian Artillery) followed by A and B Companies of the Mounted Infantry under Captain Harry Chauvel and Captain Philip Pinnock respectively. Lieutenant Alfred Adie carried the handmade flag presented to the contingent by the women of Brisbane. The men embarked on the *Cornwall* to the tune of "Soldiers of the Queen", played by the Headquarters Band. The words and music were "catchy" and easy to remember:

> *It's the soldiers of the Queen, my lads,*
> *who've been, my lads, who've seen, my lads,*
> *in the fight for England's glory lads,*
> *Of its world-wide glory let us sing.*
>
> *And when we say we've always won,*
> *And when they ask us how its done,*
> *We'll proudly point to every one*
> *Of England's soldiers of the Queen!*

By 11 December the *Cornwall* reached Port Elizabeth, South Africa. The forty-three day voyage had not been very pleasant for the men, many of whom were ill with influenza during the trip. Ordered by the British Commander on to Cape Town, the *Cornwall* arrived at Table Bay on 12 December, and troops landed at Cape Town the following day. Major Ricardo was promoted to Lieutenant Colonel on the day he landed at Cape Town.

The man who the Queenslanders and other colonists were to serve under was General Sir Redvers Buller, recently appointed Commander-in-Chief. Buller was a veteran soldier who had fought in five wars, the first Boer War from 1880 to 1881, and against the Zulu, for which he was awarded the Victoria Cross. He was a hearty, jovial commander, popular with the troops but unfortunately, not only had he never commanded large formations of troops in the field, he had assumed command after ten years of desk service.

Buller scrapped the War Office plan of an advance on the Boer republics and decided to split his force into three columns, leaving the Cape undefended. The tragic result of this blunder was "Black Week", when in December 1899 the Boers soundly defeated the British at Stormberg, Magersfontein and Colenso. The noted British historian Arthur Conan Doyle wrote:

> The week ... was the blackest one known during our generation, and the most disastrous for British arms during the century. We had in a short space of seven days lost, beyond all extenuation or excuse, three separate actions. The total loss amounted to about three thousand men and twelve guns, while the indirect effects in the way of loss of prestige to ourselves and increased confidence and more numerous recruits to our enemy were incalculable.[4]

The pride of the British Army had been dragged in the dust of the veldt by Boer farmers, who were no longer considered simply "native rebels". In addition Kimberley, Mafeking and Ladysmith were still under siege. Buller was subsequently replaced by Lord Roberts.

The British were out of their depth as they were confronted by an enemy which used unconventional tactics. The Boers had no standing army and wore no standard uniform. Every farmer between sixteen and sixty was prepared to fight, providing his own horse, rifle and ammunition, and provisions for eight days on the open plain (veldt). These citizen-soldiers were organised into highly mobile guerilla units called "commandos" varying in size from several hundred to many thousands, appearing and vanishing as required.

One war correspondent described them as a "motley-looking" group of fighters, a "crowd one is apt to see in a far inland shearing shed in Australia".[5] However there was nothing unsophisticated about the Boer force. The artillery was trained by German officers and equipped with the most modern German and French guns, outranging their British counterparts. The mounted infantrymen favoured the German Mauser rifle and the Maxim machine-gun. The Boer was a skilled marksman, having gained much experience from years of living and hunting on the veldt. He would hide behind boulders or rocky hills (kopjes), picking off advancing infantrymen with ease, often using small white stones left on the plain as target-markers. As the infantry closed in, the Boers would leap on their ponies and disappear in a cloud of dust.

When Lord Roberts assumed command of all troops he found an army extended over a front of 500 miles. The task of the new Commander-in-Chief was to take the offensive by carrying the war across the borders and into the republics and, in doing so, force the Boers to loosen their grip on the besieged British towns.

The Queensland contingents, together with those from other colonies never fought as one Australian division. They were divided up, as were the Canadian and New Zealand contingents, and attached to British regiments, thus fighting in what was an Empire army. They learnt much soldiering from the British, while the British learnt a great deal from the colonials who were well-suited to guerilla warfare, and to the country itself.

After arriving in Cape Town the Queensland Mounted Infantry went immediately by train to Orange River providing some badly needed mounted troops to the Kimberley Relief Force. They saw no action for two and a half weeks, but this was still far too short a time for the horses to be acclimatised and recover from the sea voyage. On first arrival in Africa the horses developed a kind of influenza and a regiment on the march sounded like a veterinary hospital with the sneezing and wheezing of the wretched animals.

The loss in horses from starvation, disease and sheer exhaustion was terrible. Captain Chauvel was to write home early in the campaign that they were losing five horses a day. The situation only became worse; it was unlikely that a man would ride the same horse for very long, there was therefore little opportunity of building the extraordinary sympathy between rider and horse that later existed in the Middle East in 1916–18, when well-cared for horses displayed enormous stamina and great heart.

The turnover in animals was so great that by the end of the war no fewer than two hundred thousand horses and mules were lost, the carcasses lying the length and breadth of South Africa. When the war concluded there were 264,000 horses and mules and 19,000 oxen listed on army service.

> A day on the march was very much as follows – At grey dawn the soldier gave his horse a meagre (very meagre) supply of hard

uncrushed Indian corn to eat. No hay, nor bran, nor any other fodder was supplied to assist the animal to chew and digest the unattractive maize. While eating his feed, a saddle, loaded with accoutrements up to a weight of six stone was hoisted onto the horse and left there while the trooper went to get his own scanty breakfast. At midday again a very small meal (about a handful) of maize or raw oats would be given to the horse. Such water as they got to drink was hurriedly snatched at intervals of marching, and on many occasions the horses went all day without water.

At night the troopers would be out on outpost duty, and this meant the horses were kept standing all night with their saddles on, unable to rest. Many a time the horses went 48 hours without having their saddles off; and there was no chance for them to recuperate. Day after day they had to plod on under the blazing sun all day, and in the freezing wind at night.[6]

The few cavalry charges that were attempted during the war were sorry spectacles; a long drawn out string of weak and weary horses, plodding hopelessly across the veldt at a canter, urged to further exertions by blows from the riders, and with no hope of closing on the enemy. All of the combat was of the nature of mounted infantry in which the men dismounted to fight.

Sunnyside – The First Encounter

The first opportunity for the Queensland contingent to distinguish itself as something more than a scouting body came on New Year's Day 1900, at Sunnyside.

On 31 December a flying column left the camp at Belmont. In an effort to conceal its real objective, the column veered east from the railway before recrossing the line and camping at Thornhill, west of the railway. At Thornhill the column commander, Colonel T.D. Pilcher of the Northumberland Fusiliers, addressed all ranks to make known the objective. They were to attack a Boer defensive position (laager) in the Sunnyside kopjes, about sixty miles southwest of Kimberley.

The column was composed of Queensland Mounted Infantry

3 South Africa, the area in which the Queenslanders were employed during the Boer War.

with their machine-guns, one hundred Canadian Infantry, two guns of the Royal Field Artillery and a Maxim gun. In light early-morning rain the little command dismounted and began carefully working up the east side of the kopje, drawing fire but also returning fire to keep the Boer's heads down.

Colonel Pilcher also sent out a couple of four-man flank patrols so that his advance and attack would not be caught in a crossfire. Lieutenant Adie led one of these patrols consisting of four men — Privates Herman, Butler, Rose, and Jones — springing just such a Boer trap. While riding along in open country, the patrol saw four Boers riding in on their right. Lieutenant Adie ordered the Boers to surrender; straightaway twelve more appeared. The Boers opened fire and Private Victor Jones from Mount Morgan was shot through the head and died instantly. Private Victor Jones had received pre-war training with the Quensland Mounted Infantry. He was an employee of the Mount Morgan Mining Company in Central Queensland where he had started work as an office boy and at the time of his enlistment held the position of paymaster. During the fighting Lieutenant Adie was shot twice and his horse killed from under him. Seeing the danger their commander was in, Privates Butler and Rose dashed to Lieutenant Adie's rescue. As they were carrying their leader away, the Boers succeeded in wounding Private Rose and killing his horse. All but one of the little force had been shot and three of the horses had been lost, yet they had prevented a Boer encirclement. They struggled back to the main body in time to join in the main attack.

Shortly after 11a.m. the first shot was fired, and shooting continued streadily until 1p.m. when Colonel Ricardo, with Captain Chauvel's A Company, led the Regiment in a direct attack. Taking advantage of every bit of rock and cover, moving slowly onwards, shooting only when they saw their mark, the Company moved in on the Boers. The Boers gradually retired over the hill.

Meanwhile B Company, under the command of Captain Richard Dowse, had worked its way around the left flank. As the two companies moved in they completely surrounded the Boers, catching them in a trap. Soon after 3p.m. the Boers hoisted a white flag. A total of six Boers had been killed, twelve wounded, and forty captured together with a large quantity of rifles, ammunition and equipment. The ammunition was burnt and their rifles smashed against the rocks. It was during the attack Private David McLeod was killed when shot through the spine. Private McLeod served as a gunner in the Queensland Permanent Artillery. In Victoria on a training course when the Queensland contingent was being raised, he had returned immediately to Brisbane to enlist.

As the Queenslanders were sent in pursuit of the Boers, the Canadians conducted the sad ceremony of burying Private McLeod. A grave was dug and a New Testament found. Major Bayly, the Staff Officer of the expedition, read a few selections from Corinthians over the body, after which it was consigned to the veldt. A crude cross bearing his name and unit was erected to mark the spot. Private Jones received a less formal burial. On the following day the Queenslanders conducted a search to find his body, having found it they buried him without ceremony or rite, but with deep feelings of sorrow.

The fight on New Year's Day was a small affair, yet it was at Sunnyside that two Queenslanders, Privates McLeod and Jones, became the first Australian soldiers to die in battle. In a letter home a Queensland soldier said, "It was the 1st January 1900 – a day and a year to be remembered. Queensland has had the honour of losing the first blood shed by the Australian colonies in the defence of Britain's rights."[7] Captain Chauvel later had a monument erected at the historic spot. This small monument, bearing the crest and motto of the Queensland Mounted Infantry, was the first Australian military memorial erected on a battlefield. However, the memorial deteriorated over time and some years ago the ruined memorial was removed to Kimberley where the stones were crushed and mixed with the mortar used in binding the permanent British Memorial erected in the Garden of Remembrance by the War Graves Commission.

Almost 90 years later, at dawn on Anzac Day 1989, a replica of the Sunnyside memorial was unveiled at Enoggera Barracks by the Honorary Colonel, Brigadier C.D.F. Wilson. The replica occupies pride of place in Chauvel Drive in front of the regimental headquarters of the 2nd/14th Light Horse (Queensland

Mounted Infantry), where it serves as a reminder of sacrifices made by former members of the Regiment.

Sunnyside was the first success since "Black Week"; an excellent start to the new year, it served notice to the Boers that here was a different kind of adversary, one that demanded greater respect than that shown to the British. News of the victory was reported throughout the Empire and when the Brisbane newspapers published telegraphic accounts on 3 January, there was a dramatic increase in the number of volunteers to join the second contingent, then being formed at Meeandah, Queensland. Of the Sunnyside episode a correspondent of the London *Daily Mail* wrote:

> The picturesque figures of the expedition were the Australians, whose very postures and loose costumes reminded us of cowboys brought to London by Buffalo Bill. All of them wore wide-brimmed hats. In some cases the brims of the hats were turned up, in others not. In some cases the crowns were punched in, in others they were shaped like sugar loaves. No two men carried their carbines in the same manner.
>
> The British soldier speaks in wonder of the amazing quickness of the Australian in mastering the country. The Colonials can find their way in the darkest night of any district in which they have gone by day, and every man can fight on his own account without having to be officered.[8]

On the night of 1 January the Queenslanders and Canadians spent a well-fed evening camped at a Boer farm from which they looted twelve goats, one hundred fowl and a large quantity of good hay for the horses. Throughout the campaign the Australians helped themselves to "supplementary" rations, although this was strictly forbidden. The Canadians so the story goes, were also guilty of looting, an activity which only increased the hatred and animosity between the Boers and the British.

On 1 February, the Queenslanders changed camp to Rietfontein some miles south of Belmont. Part of the contingent went out from here on an expedition with Lieutenant Colonel Boyd Royal Engineers. The trip was a pleasant one. It gave both men and horses the treat of plenty of fresh spring water and a rare chance to bathe in a large dam. Late on the evening of 9 February, orders came through that the men of Colonel Boyd's expedition were to march at midnight. Tents were struck immediately, and they returned to Belmont, reaching it around 3.00a.m.

Many horses had not withstood the strain of the meagre diet and the lack of time to acclimatise – a great number had already been lost. Those who still had horses in fair condition set off a few hours later for an undisclosed destination. After some distance they found themselves part of about ten thousand horsemen – cavalry, mounted infantry and artillery – on the move across the veldt. At dusk they arrived at Ramdam where they joined a column of some twenty thousand infantry acting as a guard to a huge convoy of wagons. By 3a.m. the following day the whole enormous army, under the command of General Roberts, commenced the march towards Kimberley.

The Relief of Kimberley

The march involved the fitting out and organising of thousands of wagons drawn by tens of thousands of oxen and mules. A remarkable feat of logistics. General Roberts commenced the advance in February, the hottest month of the year when daily temperatures on the veldt often exceeded one hundred degrees fahrenheit. On the first day the army had to cross sixteen miles of waterless veldt and both men and horses suffered. The infantry fared the worst with nearly two thousand men out of the four thousand strong 15th Brigade overcome by heat and thirst.

Major General French, a lifetime cavalryman, was given a double brigade with which to keep the Boer horsemen, led by General De la Rey and General Christiaan de Wet, occupied during the advance to Kimberley. French's command included the New South Wales Lancers, the Queensland and New South Wales Mounted Infantry, and two troops of the Australian Horse. The Australian's emu plumes waved alongside the helmets of historic British regiments such as the Inniskilling Dragoons and the Scots Greys.

The Boer General Cronje was unconcerned that General French was on the move as he believed that the British could not operate at any distance from the railway. When he realised the infantry was following up the mounted troops he sent nine hundred men to block the route near the Klip Drift. From the Drift, rising ground led to the Boer positions on both sides of the saddle (nek). Quickly summing up the situation, French ordered a cavalry charge through the centre of the nek. In extended formation under covering artillery fire, six thousand horsemen galloped up the slope in the face of both converging and frontal fire, in what was one of the few successful charges of the war. The speed of the charge through the moving wall of dust made Boer fire ineffective. At the sight of the charging squadrons bearing down on their thin line, the Boers fled. The door to Kimberley was open.

On 15 February 1900, General French entered Kimberley. Thus ended a siege which had lasted four months and had seen over forty-eight thousand people sheltering within a perimeter of some twenty miles. "At Kimberley", wrote Banjo Paterson, "the people simply hurled themselves at the horses and cried and wept for joy."[9] The next day the Queenslanders and New Zealanders took part in an action at Dronfield, eleven miles to the north of Kimberley. Captain Chauvel wrote, "We had a pretty hard day's fighting the day after we arrived driving the Boers out from around Dronfield."[10]

However all had not gone according to plan. A commando led by General de Wet attacked and destroyed a convoy, capturing over two hundred wagons, these contained eight days food for the troops and forage for the horses. The loss was felt in the weeks ahead when the entire army, both men and horses went to half-rations. The rationing continued until Bloemfontein was captured.

Battle of Paardeberg Drift

Following the relief of Kimberley, Lord Robert's army discovered General Cronje's position on the morning of 16

February. The column commander, General Kitchener despatched his mounted infantry in hard pursuit with orders to follow and stick with the convoy at all costs. Dawn on the 18th saw the Boers in a laager concealed along two miles of riverbed. Between the high banks and shallow waters of the river they found room for their cattle and horses. During the night and following days the Boers dug furiously into the banks and constructed shelter for their women and children. They cut slit trenches along both banks and occupied the dry creek beds that led to the river.

Kitchener ordered an immediate assault, anxious to take the laager before Boer reinforcements could arrive. At 6.00a.m. on 18 February the main attack by the 13th Infantry Brigade and Highland Brigade was launched over 1500 yards of bare ground which offered the assaulting troops the minimum cover. The attack faltered 500 yards from the drift. The men clung to the earth in the sweltering heat for nearly twelve hours before sneaking away under the cover of darkness. Towards noon along the northern bank just above Paardeberg Drift another assault was launched by the British and Royal Canadian Infantry. By early afternoon this assault also ground to a halt. Kitchener ordered all men lying within 500 yards of the Boer positions to charge with the bayonet. When the troops rose to their feet they were decimated by heavy fire from the Mausers.

The long and torrid opening day of the battle had proven a severe test of stamina and courage. The infantry had attacked across an open field in daylight against a well-entrenched enemy equipped with the most modern high velocity magazine-fed rifles. The British and colonial armies had suffered the highest casualties of any single day of the war with 320 killed and 1,392 wounded, including a number of Australians attached to British regiments.

Lord Roberts arrived at the position on 19 February and established a cordon, approximately four miles in diameter, around the Boers. The cordon, including the Queensland Mounted Infantry, succeeded in containing General Cronje's army. It also intercepted and repelled 3,500 Boer reinforcements arriving from the east. The cordon held although General de Wet with 600 men succeeded in occupying a position overlooking the British positions.

When Lord Roberts discovered that women and children were within the laager, he offered General Cronje safe conduct through his lines for all non-combatants. He also offered medicines and surgeons to treat the wounded. General Cronje rejected outright the offer of safe conduct and would only accept medical assistance if the surgeons remained in the laager until he was ready to leave. Lord Roberts rejected this proposal.

With the benefit of an observer in a balloon, the British now pounded the Boer entrenchment with some 91 artillery pieces including 4.7 inch naval guns. Little movement was now possible within the confines of the river banks. On 23 February it started to rain, falling increasingly for over twenty-four hours. The problems of the besieged Boers increased when the river rose by several feet. Still they refused to surrender! Under the cover of darkness the British engineers, with infantry support, built trenches not less than one hundred yards from the Boer positions. Finally on 27 February, when the sun rose the Boers found that their positions were untenable and at 7.00a.m. General Cronje ran up the white flag. In all, more than four thousand Boers surrendered, handing over large quantities of weapons, ammunition and supplies. Over 170 wounded were extracted from the trenches where they had been lying without medical assistance. The trenches were a marvel of military engineering, having withstood a continual artillery bombardment for ten days.

Bloemfontein

Although anxious to march on Bloemfontein, Lord Roberts kept the army at Paardeberg for a week after the surrender of Cronje because of the poor condition of the horses and the now boggy roads. When the move started the army had a strength of 30,000 men, 12,000 horses, 10,000 mules, and 116 guns. The men and horses were both reduced to half-rations. The Boers withdrew before the advancing army although there were several fierce clashes.

Two miles from Bloemfontein the Mayor and officials presented Lord Roberts with the keys of the town. For the soldiers a long and tiring march had ended. They had marched for a month without tents and on short rations. The tired and worn troops had left lying along the way no less than two thousand carcasses of horses that had died of exhaustion and starvation.

Private Norman Seccombe of the Queensland Mounted Infantry (QMI) described the march: "Advancing at the rate of about 10 miles a day, we were cut to half-rations, three biscuits a day — very hard — half a pound of tinned meat and a quart of tea. The poor wretches, horses and mules, that die by the wayside would soften the heart of the hardest."[11] Corporal J.H.M. Abbott, serving with the Australian Horse wrote: "The day we captured Bloemfontein I had nothing to eat for 30 hours but half a corn cob and a biscuit."[12]

Sanna's Post

Following a period of recuperation at a rest camp at Springfontein, the Regiment was involved in the relief at Sanna's Post. From Bloemfontein the British right flank extended some forty miles to Thaba 'Nchu, a small town garrisoned with a force commanded by Brigadier Broadwood. Under increasing Boer pressure Broadwood commenced a withdrawal towards Bloemfontein on 30 March. Broadwood's convoy consisted of stores, and families of loyal citizens from the district protected by a force of eighteen hundred men and two batteries of Royal Horse Artillery. In a well-executed ambush the Boers, under the command of de Wet, inflicted heavy casualties — British losses were 159 killed and wounded; with de Wet capturing 421 prisoners, 7 guns and 93 wagons loaded with stores.

A relief force was ordered to assist Broadwood's troops. The force, which left Bloemfontein on 31 March, included two companies of the 1st Queensland Contingent under Lieutenant Colonel Ricardo and thirty men of the 2nd Contingent under Captain William Thompson.

On arriving at the battle, Captain Thompson and the men of the 2nd Contingent were detached to guard a transport train, while Ricardo's force was ordered to reinforce a kopje that was being held by the Southern Counties Mounted Infantry. As they reached the kopje Ricardo's men were met with heavy shell and rifle fire which killed several horses, but the men escaped unscathed. Ricardo ordered Captain Richard Dowse and about forty men to occupy a small fold in the ground on the left, which was being threatened by the enemy, and from which the kopje could be enfiladed.

Captain Dowse had to cross a river to reach the support position but on arrival found it not to be ideally suitable, so he pushed on another 800 yards to a fold in the ground from where any approach by the enemy could be effectively stopped. Although the area was continually swept by heavy rifle fire, Dowse and his men gallantly held their position for more than five hours against an enemy force estimated at over one thousand. Time after time the Boers tried to advance and drive them out. At one stage the position was threatened with envelopment, but Lieutenant Adie (who had recovered from the wounds received at Sunnyside) with twenty men stopped the flanking movement of the Boers.

One of the NCOs present during the battle expressed his admiration,

> It would be impossible to picture a braver man than Captain Dowse. No matter how black things looked — and they looked black enough once or twice — his splendid example steadied and encouraged us. He was perfectly cool, and nothing escaped him. When poor Conley was shot, Captain Dowse and Corporal Stevens crawled over to him, and our brave commander bandaged up the wound with skill and tenderness. And when it came to the retirement, and bullets were flying round us like hail, Captain Dowse stopped and went over to the assistance of a wounded man. He was always very strict; some of us thought he was too severe, but we would never ask to serve under a braver officer or one who can handle his men better in action.[13]

At 3.00p.m. Ricardo received the order to withdraw; he sent his men back in small groups each by a different route, and as each little band galloped off it became the target for artillery and rifle fire. Ricardo's men escaped without loss, but nineteen horses were killed. Captain Dowse and his men provided protection for Ricardo's withdrawal, any premature move on their part would have led to the capture or death of Ricardo's men and the Imperial Mounted Infantry.

It then came the turn of Dowse to retire and he had a difficult task before him. Several hundred Boers were now firing at his little command. Dowse sent Sergeant Major Richard Breydon and ten men forward to hold one of the drifts while he ordered the others to scatter and gallop away to a rendezvous on the other side of the river. During the withdrawal Private Charles Strong of Chelmer was seriously wounded, but prompt action by Captain (Surgeon) Joseph Dods saved his life. Sergeant Major Breydon and his men suffered at the hands of the Boers, Private Herbert Reece being killed by a bullet through the head.

Once the company was clear of the position, the Sergeant Major and his men made a dash for their lives. By this time they were being pursued by over 400 Boers and being fired at by rifles, machine-guns and artillery. They had about 400 yards to ride to reach safety, but their route was blocked by a creek. As they charged into the water the horses sank to their saddle flaps in the mud. Within seconds the Boers surrounded the floundering horses and men. Sergeant Major Breydon and four of his men were marched off into captivity.

The Fall of Pretoria

By the end of April sufficient supplies had been stockpiled to commence the next phase. Remounts had arrived, wagons repaired, teams of mules and oxen gathered, and the army had been reinforced and rested. Lord Roberts believed that the fall of Pretoria would herald the end of the Boer resistance. Strategically the capital was important because of the necessity to seize the railway at Delagoa Bay, the only sea link open to the Republics from which they could draw military supplies and maintain overseas contacts.

The force, under the command of Lord Roberts, totalled

24,000 men and 210 guns with a further 14,000 men and 48 guns moving parallel to Roberts' force under General Hamilton. Opposing these armies the Boers could muster no more than fifteen thousand men under General Botha. For the great advance all of the Australian contingents, numbering a little over three thousand mounted soldiers, formed the spearhead of the army, riding ahead to reconnoitre or scout to the flanks. The Queensland Mounted Infantry departed Bloemfontein on 1 May as the general advance to Pretoria began.

The Queensland Mounted Infantry took part in actions at Brandefeit, Vet River, Zand River, Klipriviersberg, and Johannesburg. On 31 May, Johannesburg was surrendered to Roberts. That same afternoon the Queensland Mounted Infantry, when crossing the Rand River to the north of Johannesburg, spotted some five miles away a small commando, with guns and wagons, slipping north towards Pretoria. They gave chase and by nightfall had captured Commandant Runck of the German Legion, a 3 inch gun and gun wagon, wagons loaded with military stores and ammunition and twenty-three prisoners.

In Pretoria the Boer government decided against making a firm defence of the town. On 30 May, President Kruger departed for Middleburg on the Delagoa Bay railway, taking with him the state archives and funds. He was later to flee to Europe aboard a Dutch warship. General Botha received instructions to delay the occupation for as long as possible while avoiding a serious engagement. Lord Roberts' army was now over a thousand miles from the main base at Cape Town and had suffered considerable losses of men, horses and equipment in three months of fighting. Instead of waiting for reinforcements and supplies to arrive, he acted decisively and ordered his men to pursue the Boers towards Pretoria, some thirty-five miles from Johannesburg. His speedy advance resulted in Pretoria surrendering on 5 June.

Boers Cut Supplies

Although the Boers had been frequently dispersed in the field, they were far from beaten. The problem for Roberts still remained of how to bring them to a decisive battle. The rapid advance to Pretoria had resulted in an extended line of communication with very few troops available to protect the hundreds of miles of railway lines and roads. General de Wet made the most of the opportunity by cutting the track, destroying culverts, seizing convoys, and tearing up several miles of railway line. The result was the loss of winter clothing and soldiers' mail, with Pretoria temporarily isolated.

After the occupation of Pretoria, strong inducements were offered to the colonial troops to transfer to the South African Constabulary, then being raised. A considerable number of soldiers accepted the offer, reducing the strength of the Queensland contingents. As the Queensland Mounted Infantry was now reduced to two companies, Lieutenant Colonel Ricardo handed over command to Major Chauvel. These companies were engaged in operations at Reit Vlei, the advance to Balmoral, the fighting at Zilikat's Nek, and Olifant's Nek during July and August.

In the coming month General Hunter succeeded in blocking a number of neks in hilly country resulting in the entrapment of a large Boer force. On 30 July more than four thousand burghers laid down their arms and surrendered unconditionally. In addition four thousand horses and a million rounds of ammunition were seized. This was the second great defeat for the Boers and over eight thousand were now prisoners of war, a significant proportion of their available manpower.

Orders were received for the return of the 1st Contingent to Australia, and in early November the saddles and horses were handed over and the men entrained for Cape Town. On 13 December 1900, the anniversary of their disembarkation, the 1st Contingent boarded the transport ship *Orient*. They arrived back in Brisbane on 17 January and disbanded on 23 January.

Relief of Mafeking

Throughout 1900 men of the 2nd and 3rd Contingents continued operations, including the relief of Mafeking. On 5 May a squadron of the 3rd (Queensland Mounted Infantry) Contingent, commanded by Captain Charles Kellie was detailed to escort the Canadian artillery in the relief of Mafeking. Because no horses were available for the first leg to Bulawayo, stagecoaches drawn by mules were used. Over the last thirty miles the mules were replaced by bullocks. The escort for the guns, piled up inside and on top of the coaches, arrived in Bulawayo after a journey of 280 miles. The squadron then travelled the next 460 miles by train to a point some 4 miles north of Mafeking where the Boers had destroyed the track. Still without horses, they trudged forty miles on foot to join the column at daybreak on 14 May 1900, at the end of a long night march.

Seven hours after joining the relief force the Queenslanders marched to a point twenty miles west of Mafeking to join forces with a flying column from Kimberley. On 17 May this force met and defeated the Boers in an action lasting four hours. Being tasked to protect the wagons, the Queenslanders took no part in the action until late in the day. The infantry, veterans of previous campaigns, proceeded to attack. Fearing that they would miss out on the action the Queenslanders left the wagons and rushed forward to join the infantry, cheering and yelling as they did so! Soon they found themselves running stride for stride with the British infantry to the Boer positions.

The Premier of Queensland received a telegram from Lieutenant Colonel R.S.S. Baden-Powell dated 17 May and sent to Kimberley by runner, "Mafeking relieved today. Am most grateful for invaluable assistance by Queenslanders under Kellie, who made record march through Rhodesia to help us."[14]

Following the capture of Pretoria and relief of Mafeking, garrisons were established along the Mafeking to Rustenburg road to provide staging posts for the convoys. Altogether more than two thousand men were spread out along several hundred miles of road. However there were still more than seven thousand Boers operating between Mafeking and Pretoria and it was not long before they started to interdict the supply route. In mid-July the Boer General Lemmer occupied a position of strength near the Koster River, halting all convoys.

Koster River

With the Boers blocking convoys between Zeerust and Rustenberg the supply situation became serious. Baden-Powell ordered a force, under the command of Colonel Airey to "brush aside" the enemy, proceed to Elands River Post, and return with a stranded convoy. Late on the afternoon of 21 July, a force of 300 Australians, including the Queenslanders, started to move along the road to Elands River. At dusk the Boers opened fire on the scouts. That night the column camped on a ridge away from the road.

Early next morning the whole force came under fire after the enemy on the nearby hills had allowed the flanking scouts to pass. Under heavy fire the Australians flung themselves onto the ground, though there was precious little cover.

Captain Richard Echlin, QMI, gave the following description of the action:

> The Queenslanders had the whole brunt of the fire, being in the lead, and first to come in view of the enemy. No time was wasted in lying down in such cover as could be found. It was sorry protection from the notorious Boer sharpshooters at 800 yards range. The only reason I have for every man not being shot in the first minute or two was that the Boers must have directed their fire on the horses as they were being led to cover. This gave our men a little time to extend and pick their positions.
>
> The cover was best in the long grass near the road. There was not a single stone or rock, but some trees six to ten inches in diameter, and some shrubs not unlike saltbush. Our men replied erratically at first, but when properly set to work, never firing at random. It would never have done to blaze away their 200 rounds when we knew we must wait for relief of darkness for our salvation.[15]

All day the Australians were trapped by an enemy force of about one thousand. From a range of about eight hundred yards the Boers kept the Queenslanders trapped in the long grass alongside the road, perspiring in the sun and becoming parched from the lack of water. It was a desolate scene with stampeding horses charging through the position, and helmets, haversacks, equipment, and shot horses scattered all over the place. Relief was finally summoned by Miss Bach, a young Englishwoman who lived in a farmhouse nearby. At 2.30p.m. volleys of shots could be heard from the nearby kopjes as the relieving force despatched the Boers. "It was the first time any of us could stand at full length without the certainty of being shot since 8a.m."[16]

In the battle of Koster River the Australian casualties amounted to thirty-nine, including nine killed. They are buried in the town cemetery at Rustenberg.

Outstanding Bravery

Early in the battle Bugler Arthur Forbes, an eighteen year old youth with QMI was given the task of horseholder. He moved the horses to a sheltered position behind a deserted farmhouse. The Boers soon noted their presence and in no time the horses were shot, one bullet passing through the bugler's haversack. Forbes retired with other horseholders into the farm in which a number of Australians were making a stand by firing through holes knocked in the walls. When ammunition ran low, Forbes went out several times under heavy fire to the dead horses and salvaged cartridges from the saddle wallets.

For outstanding courage Bugler Forbes of Milton was mentioned in despatches and awarded the Distinguished Conduct Medal. After the war the Governor-General of Australia presented young Forbes with a silver bugle and a purse of sovereigns. In the 1914-18 war Forbes served as a chaplain, then as a gunner in the artillery. In July 1917 he became a chaplain again in the AIF serving in France and England. His appointment to the AIF was terminated in 1944 when he was placed on the retired list.

Elands River

Compared with the importance of the siege of Ladysmith, or the investiture of Kimberley and Mafeking, the trapping of a small garrison at Elands River in August 1900 seems but a minor affair.

But the war furnished few more notable instances of brilliant courage and endurance. The garrison consisted of Australian and Rhodesian irregulars, under the command of Lieutenant Colonel Hore of the 5th Dragoon Guards, who had previously commanded the Protectorate Regiment at Mafeking.

The post at Elands River, garrisoned by a small detachment of Rhodesians was considerably reinforced with troops. The post held a large quantity of stores, stockpiled due to the disruption of the line of communication between Mafeking and Rustenburg. By early August, Lieutenant Colonel Hore had under his control: 140 men from the Queensland Mounted Infantry commanded by Major Walter Tunbridge, 100 New South Wales Citizen Bushmen, 40 Victorian Bushmen, 9 from West Australia and 2 from Tasmania. In addition there were 200 troopers of the Rhodesian Volunteers and a handful of Canadian and British horsemen. Other than their personal weapons they had an old muzzle-loading 7 pounder and two Maxim machine-guns.

The force held a low hill, strewn with boulders, in the centre of a small plain. The river was half a mile to the west, with two small hills beside it. The Boers moved into a position on the hills overlooking the Australian force. They were able to creep up to within rifle range by using the boulders as cover. The Bushmen in the ranks had little appreciation of the danger of their position holding such an amount of stores in the midst of an enemy lacking supplies. Although intelligence was received that the enemy planned to attack the post, the men could not be persuaded to take a keen interest in digging trenches in the hard surface.

On 3 August a message was received by telegraph stating that General Carrington, with a force of over one thousand men, six field guns and four pom-poms, was expected to arrive on the fifth to cover the withdrawal of the whole garrison to Mafeking. That night the camp celebrated with a cheerful campfire concert, the lights of the fires and the sound of the singing carrying to the surrounding hills. Only a handful of officers knew that the Boer General, De la Rey, with one thousand burghers and accompanied by artillery, lay in the nearby hills ready to strike at any time.

Early on the morning of the 4 August, the first alarm came

when the Boers opened fire on the watering parties undertaking the daily task of collecting water from the drift, approximately a half a mile from the camp. The Boers then commenced to shell the centre of the camp. Their fire was accurate, destroying the telegraph communications from the camp to the outside world and falling on the horse lines. In no time the area was a shambles. The animals, estimated at over fifteen hundred, were on an exposed slope and were bowled over like ninepins. The frightened and badly mutilated animals stampeded in terror, some struggling on shattered stumps of legs, others, with mangled bodies, stumbled into trenches. On that first day, when over seventeen hundred artillery shells fell on the camp, the slopes literally ran red with blood. Under the cover of darkness, in an effort to clean up, the surviving mules were used to drag away the carcasses.

The camp also came under accurate fire from pom-poms, shrapnel, 12 pounders, Maxims, and heavy rifle fire. The fire came from all directions. The big guns of the Boers were out of range of the sole 7 pounder in the camp, which was of little use to the defenders, although they managed to score a direct hit on a farmhouse from which snipers were operating. Major Tunbridge worked untiringly under fire, to try and keep the gun in service; four times it had to be dismantled to effect repairs. Tunbridge spent a day and a night with a file repairing some of the shells, many of which had been damaged in transit.

On the first day the defenders suffered thirty-two casualties. The wounded were attended to in the three ambulance wagons which were to serve as the makeshift hospital during the siege. The hospital was situated near the centre of the camp close to the horse lines, its only protection a double wall of biscuit tins covered with tarpaulin. Private John Masterton, the first Queenslander hit, later died in a hospital at Krugersdorp. Surgeon-General Albert Duka of the QMI did splendid work under the most atrocious of conditions: initially, construction of the hospital was not complete and operations had to be performed out in the open. Surgeon-General Duka was afterwards awarded the DSO for his efforts.

When darkness descended all who could be spared for the task

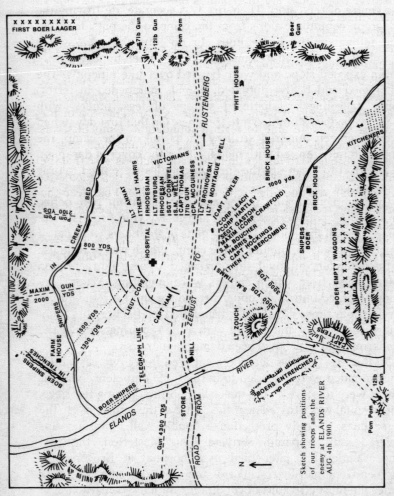

4 Battle of Elands River. Plan drawn from sketches by Corporal Kelman and Trooper Donkin, of Major Tunbridge's forces, further embellished by Captain Cope and Chaplain Green, all of

of digging worked like beavers until dawn. They needed no persuading now! With digging tools in short supply they used whatever they could lay their hands on – including bayonets. Although no large stones covered the area, rocky slate protruded through the surface in a number of places. The men dug, gouged and levered to raise the slate slabs to the surface in an effort to get under cover by dawn. Without the use of entrenching tools and defence stores, everyone showed his initiative utilising whatever resources were available. By dismantling wagons and placing the wheels or parts of the chassis across the top of the trenches they were able to form a foundation. Onto this was placed an overlay of earth or slate, meat cases and flour bags – smeared with mud to render them less conspicuous – so that the trenches became almost bombproof. Because the defended area was so small, wagons filled with earth were placed between the backs of the trenches to try and eliminate the danger of being shot in the back by cross-fire.

Lieutenant James Annat led a patrol of twenty-five Queenslanders in an effort to silence a particularly troublesome pom-pom. By crawling through the grass for more than two hundred yards the patrol opened fire so effectively that the Boers were forced to retire. Later Annat unsuccessfully sought permission to take a raiding party out at night to try and capture the gun. Lieutenant Annat had taken part in the relief of Mafeking. He served with the distinction in the early days of the siege and he often went out into enemy-held territory for hours to signal back the range of the Boer guns. At dusk on Monday 6 August he was killed by a shell exploding at his feet. That night his men carried his body, covered by a Union Jack, to a burial spot just outside the trenches.

Early on the second day of the siege, Sunday 5 August, the Boers could be seen through field-glasses, preparing for the arrival of General Carrington's relief force. They moved several guns and one hundred riflemen into position. They were observed marking the range on the road for use by the artillery and riflemen. The relief force advanced to within two miles of Elands River before being engaged by enemy artillery. The men continued to ride forward until they were fired upon by the Boer riflemen waiting in well-sited positions on both flanks. At this stage Carrington's scouts could actually be seen from the beleaguered camp. The Reverend James Green told how, by using the holes in the walls around the hospital, he had observed the action through field-glasses and described its progress to the wounded.

Soon after 4.00p.m. the Boer artillery was able to land a salvo near where General Carrington's staff had halted. One shell fell between the General and Lieutenant Colonel Wallack, a Tasmanian, killing several horses. This fire, coupled with severe casualties among the horses in the gun teams, caused Carrington to order a retirement to Marico River some seventeen miles back. Although hounded by snipers all the way the force withdrew in sound order.

On 6 August Baden-Powell marched with a force of over one thousand men. He had orders to relieve the Elands River Post. As he approached the camp the sound of fighting could be heard growing fainter. Baden-Powell was convinced that Carrington had relieved the camp and that the whole force was being evacuated westward towards Zeerust. He turned back taking his column towards Pretoria. The Australians were now on their own.

Carrington arrived in Zeerust convinced that the beleaguered colonials could not possibly hold out much longer. He telegraphed Lord Roberts that the men at Elands River had no option but to surrender. On 11 August the *Sydney Morning Herald* reported that the "Boers had barred Carrington's relief force and 300 Bushmen were captured". There was public concern in Australia and more detail was requested from the British High Commissioner at Cape Town. His reply, referring to a report issued by Lord Roberts, was, "The garrison at Elands River has I fear been captured".[17]

But captured they were not! Surrounded, outnumbered and outgunned though they were, in their isolated post. The siege soon attracted the attention of several small commandos, so that the attacking force grew to somewhere between two and three thousand. In anticipation of hauling off a great amount of stores the Boers assembled a long line of wagons.

On 8 August, the Boer commander, De la Rey sent a messenger under a flag of truce to advise Lieutenant Colonel Hore that the relief forces had withdrawn and to let them know that the whole area was in Boer hands. He offered to escort the force to the nearest British post provided that none of the supplies within the camp were destroyed. He concluded, "Your commissioned officers, in such a case, will retain their arms in recognition of your courage in defence of your camp."[18] After consulting with Major Tunbridge, the offer to surrender was rejected by Hore. On 12 August, De la Rey sent a second offer of honourable surrender to which Hore replied, "Even if I wished to surrender to you – and I don't – I am commanding Australians who would cut my throat if I accepted your terms".[19]

Realising that the quickest way to force the submission of the garrison rested in stopping the water supply from the drift, each night there were skirmishes between the water parties and the Boers. As a result barely sufficient water was obtained to sustain the force from one day to the next. The Boers attempted to capture the kopjes near the river, held by Captain Butters and Lieutenant Zouch, and which provided much needed protection for the water-carts and their teams.

After a day's bombardment the Boers attacked the south kopje, held by Butters with eighty Rhodesians and New South Welshmen, storming up the hill at dusk. The colonials held their fire until the assault was within fifty yards of the trenches, and repelled the attack with concentrated fire from both kopjes and the machine-gun. On the following day the enemy tried again, this time going up the slope behind a herd of sheep and goats. Captain Butters, who had experienced such deception in the Matabele Wars, was not deceived. The sheep, goats and Boers were put to flight by well-timed volleys from the kopjes.

Although they had the numbers, the Boers, unwilling to accept the heavy losses it would cost, never attempted to storm the camp. The snipers continued to harass the camp. Although night parties had been forbidden by Lieutenant Colonel Hore, a number of Australians ignored this order. One night a Bushman, after removing his boots and pulling on four pairs of socks, crawled out of the camp to where a troublesome sniper was operating. He found the sniper asleep in the fork of a tree. The Bushman shot the Boer, then quietly returned with fifty English sovereigns found in the dead man's pocket.

Plagued by the stench from the putrid bodies of fourteen hundred animals, unwashed due to the meagre ration of a quart of water per man per day, scorched by the sun during the day and shivering at night, the garrison kept fighting. They kept the Union Jack flying, although the Boers frequently shot it down.

On 13 August the British learnt from two separate sources that the force at Elands was still fighting. A messenger from Lieutenant Colonel Hore managed to ride through the Boer lines to Mafeking. In addition, a Boer messenger was captured by a British patrol and revealed that De la Rey was still engaging the camp. On 15 August, to the surprise of the camp, the rifle fire petered out. Later in the afternoon figures could be seen some miles from the camp, but attempts to contact them failed. The next day Kitchener and his force rode into the battered position.

The men of the garrison, unable to wash for nearly two weeks, were black with dirt. All were heavily bearded, their clothes in tatters and some were without tunics. When Kichener inspected the defences he was surprised to learn that most of the trench work had been dug with bayonets. He remarked, "Only colonials could have held out and survived in such impossible circumstances."[20] On the following day the troops mustered for a service at the small cemetery situated just beyond the trenches. The mates of the fallen men made headstones of the smoothest slabs of slate from the trench parapets, carving the names so deep that they can be read to this day. The QMI formed a firing party and the Last Post sounded.

They had been outnumbered by four or five to one, they had been massively out-gunned but had not shown the slightest sound of surrendering. They had lost heavily in horses, over 1,400 of the 1,550 in the post were killed. Among the men the casualties had been amazingly light; of the seventy-five wounded only five had died. Describing the battle at Elands River, a Boer wrote:

> For the first time in the war we were fighting men who used our own tactics against us. They were Australian volunteers and although

small in number we could not take their position. They were the only troops who could scout into our lines at night and kill our sentries. Our men admitted that the Australians were more formidable and far more dangerous than any British troops.[21]

The action at Elands River was one of the greatest military achievements by Australians in South Africa. Writing shortly after the event, the British historian Sir Arthur Conan Doyle declared, "When the ballad-makers of Australia seek a subject, let them turn to Elands River, for there was no finer fighting in the war".[22] Fortunately the Australian poet George Essex Evans later took up the challenge and wrote the poem "Elands River":

> It was on the fourth of August, as five hundred of
> us lay
> In the camp at Elands River, came a shell from De
> la Rey
>   We were dreaming of home faces,
>   Of the old familiar places,
> And the gumtrees and the sunny plains five
> thousand miles away.
>   But the challenge woke and found us
>   with four thousand rifles round us;
> And death stood laughing at us at the breaking of
> the day.
> Hell belched upon our borders, and the battle had
> begun
> Our maxims jammed; we faced them with one muzzle-
> loading gun.
>   East, South and West, and Norward
>   Their shells came screaming forward
> As we threw the sconces round us with the first
> light of the sun.
>   The thin air shook with thunder
>   As they raked us fore and under,
> And the cordon closed around us, as they held us
> eight to one.
>
> We got the maxims going, and the field gun into
> place,

(She stilled the growling of a Krupp upon our Southern face);
> Round the crimson ring of battle
> Swiftly ran the deadly rattle

As our rifles searched their front lines with a desperate menace;
> Who would wish himself away
> Fighting in our ranks that day

For the glory of Australia and the honour of the race?

But our horse lines soon were shambles, and our cattle lying dead
(Where twelve guns rake two acres there is little room to tread)
> All day long we heard the drumming
> Of the Mauser bullets humming,

And at night their guns, day sighted, rained fierce havoc overhead.
> Twelve long days and nights together,
> Through the cold and bitter weather

We lay grim behind the sconces, and returned them lead for lead.

They called us to surrender, and they let their canon lag;
They offered us our freedom for the striking of the flag —
> Army stores lay there in mounds,
> Worth one hundred thousand pounds

And we lay battered round them behind mound and sconce and crag
> But we sent the answer in,
> They could take what they could win —

We hadn't came five thousand miles to fly the coward's rag.

We saw the guns of Carrington come on and fall away;

> We saw the ranks of Kitchener across the kopje
> grey —
>> For the sun was shining then
>> Upon twenty thousand men —
> And we laughed, because we knew, in spite of hell —
>> fire and delay
>> On Australia's page forever
>> We had written Elands River —
> We had written it forever and a day!²³

Scorched Earth

By September the form of the war had changed. In the former republics, which had now been proclaimed as British territory by Lord Roberts, small bands of burghers roamed. These guerillas kept out of the way of the large bodies of troops but were capable of forming into a commando of one thousand or more wherever a favourable target presented itself. A number of successful Boer leaders, Smuts, De la Rey and de Wet kept the Boer's hopes alive. Addressing the troops shortly before his departure for England, Roberts said, "As you know the war is over. There are no great battles to be won. No more large towns to be taken. The war has degenerated into a guerilla warfare, which is the most annoying part of any campaign."²⁴

Now began a much more savage campaign in which Queenslanders, as part of the 1st Australian Commonwealth Horse, were to play an unpleasant part. The plan, devised by the War Office, began with the phrase, "A war of annexation against a proud people must be a war of extermination, and that unfortunately is what it now seems we are committing ourselves to".²⁵ The plan was simple. Build a series of block-houses across the country each within rifle range of the next and run barbed wire from every block-house. The system became so successful that before long iron block-houses were being mass-produced. In the end no less than ten thousand of these small forts, each with a garrison of seven, stretched in lines over the former republics. Next the country was divided into grid squares. Then, quite deliberately one grid square at a time, the occupants were hunted out of all farms and settlements. The buildings were burnt, the livestock driven off, the wells poisoned, and the people taken to concentration camps. Troops cleared and totally despoiled a square then moved on to the next. The Boer was thus denied his local supplies of food, intelligence and hideaways at the same time he was driven towards the block-house lines. The practice of farm-burning did not meet with the approval of the colonial troops who understood the difficulty of surviving on a small farm in a hostile environment. Private Hiram Thompson of the Queensland Imperial Bushmen was to write, "We have been burning all the farmhouses, and every farm has women and children in it. We are heartily sick of the work . . . ."[26]

By August 1901 almost 100,000 whites and 25,000 blacks had been imprisoned in a total of twenty-four concentration camps. This number was exclusive of the 31,000 prisoners of war held in camps in South Africa, Bermuda, Ceylon, St Helena, and India. Conditions in the concentration camps were appalling. There was a lack of food, bedding, toilet facilities, clothing, shelter, and medical facilities. This lead to malnutrition, typhoid, measles, and dysentery. The death rate soared, with one camp at Mafeking recording 400 deaths in one month. It is estimated that a total of 20,000 white women and children and 12,000 Africans perished in the camps.

The Boers, however, continued to fight an aggressive guerilla war. On 3 December 1900 De la Rey and Smuts smashed a 138 wagon convoy drawn by eighteen hundred oxen bound for Rustenberg, inflicting casualties and capturing seventy-five prisoners. Ten days later De la Rey with 3,000 burghers launched a dawn attack on a British force near Olifant's Nek, capturing the camp stores and baggage and killing or wounding 600 men.

In the pursuit of the Boers, Australian Bushmen and New Zealanders under General Plummer were transported from Transvaal by train to Naauwpoort. Here the Bushmen went into camp with the 1st Dragoon Guards, just out from England with fine horses and eager to serve under Plummer. It wasn't long before the Dragoon's fine horses began to disappear from the horse lines, replaced by the Australians' well-worn mounts. The displeasure of the Dragoons changed to anger when they had difficulty identifying their horses again. "Give an Australian half-an-hour with a horse and tails are changed, manes are hogged, marks and brands disappear as if by magic."[27]

In drenching rain Plummer pursued de Wet forcing him to fight a rearguard action. On 14 February 1901 at Wolverkuil de Wet was forced to flee when surrounded by the Dragoon Guards, Australians and New Zealanders. Major Tunbridge, who had figured so prominently at Elands River, was mentioned in despatches by Kitchener for his part in the engagement. On the following day, when operating alone, Privates Culliford, Alford and Rule, each captured, single-handed, small bands of armed Boers. For this they were mentioned by Kitchener in his despatches and promoted to Corporal. Heavy rain delayed the pursuit, with horse teams and guns bogged on the sloppy veldt, enabling de Wet to escape once again.

The resilience of the commando leaders and their tactic of dispersing into small groups, only to reform once the danger had passed, ensured their survival. By the end of the summer of 1900–1901, none of the British efforts had resulted in the elimination of any of the leaders who inspired the Boer nation. To chase the highly mobile burghers, flying columns were introduced consisting of infantry, guns, field hospital, and wagons drawn by oxen. They were able to advance, on average, ten to fifteen miles a day. With the passing months the infantry was replaced by mounted infantry, thus greatly enhancing the mobility of the column, and the infantry were relegated to garrison duties along the lines of communication and in the towns. The Australians continued to be broken up in units all over the country.

Onverwacht

On New Year's Day 1902 three columns, under the command of General Plummer, moved out of Ermelo and travelled east with the objective of trapping Botha's commando of 750 burghers near the Swaziland border. Major Vallentin of the Somerset Light Infantry commanded one column, which comprised of companies of Buffs and Hampshire Mounted Infantry, a company of Yeoman and 110 men of the 5th Queensland Imperial Bushmen under Major Frederick Toll.

Early on 4 January Plummer's force, with Vallentin's column in the lead, marched through broken country meeting only light opposition. As the enemy withdrew, Vallentin left the Buffs to hold the main ridge at Bankkop, pending the arrival of Plummer with the main body. He went ahead for another mile then decided to halt on high ground. He posted his troops on a semicircle of hills called Onverwacht – the Yeoman and Hampshire Mounted Infantry in the centre with the Queenslanders protecting the flanks.

Acting on a report that fifty Boers were hiding in a ravine two miles ahead, Vallentin took a party forward. Shortly after starting out, his party was surprised by 300 Boers riding out of a deep hollow where they had been hiding. Outnumbered, Vallentin was driven back and made a stand on the ridge at Onverwacht. The number of Boers increased to over five hundred.

The Boers rushed into the position in an attempt to capture the column's pom-pom gun, but were stopped by the quick intervention of the Hampshires and Queenslanders. The gun withdrew and was soon in action again, but the Boers worked their way round the flank and succeeded in shooting all of the gun horses. The gun finally ended up in a gully.

Major Toll and his men, outnumbered by five to one, were forced to retire from the main ridge on foot. During their withdrawal the Boers intercepted and captured a small party. The Hampshires, a few Yeoman and the Queenslanders then made a last stand, on a bare knob of a small ridge. The enemy attacked in force and succeeded in getting within thirty yards of the position before being beaten back. Two burghers were killed less than ten yards from the front men.

Once the Boers had succeeded in working their way to the rear of the knob, the position became hopeless. The burghers swept in taking prisoners and stripping the dead and wounded of most of their clothing, including their boots. Thirty of Vallentin's horses survived the ordeal and were used by the Boers to evacuate their dead and wounded. Major Vallentin and the Boer leader, Commandant Oppermann, had both been killed.

With the approach of Plummer and the main column the Boers fled, releasing the prisoners as they would have slowed their escape. After inspecting the scene, General Plummer assured Major Toll that his men had done all that could be expected of them. The 5th Queenslanders suffered thirteen killed and seventeen wounded. Several were mentioned in despatches by General Kitchener. Bugler William Busby and Lieutenant Charles Reese were mentioned "for gallantry in action"; Sergeant James Power (who was killed) "for gallantry and good conduct in action"; Company Sergeant-Major Frank Knyvitt was awarded the DCM "for coolness and gallantry in action"; while Major Toll was commended "for the resolute and capable way in which he led his regiment". [28]

Capitulation

The block-houses, the scorched earth policy and the decimation of their women and children in the concentration camps eventually forced the Boer leaders to the negotiating table. Finally, on 31 May 1902 a surrender was signed at Vereeniging.

The final Queensland contingents returned home in August 1902. Queensland had despatched a total of 150 officers, 2,763 men and 3,209 horses to the war. Of these, 4 officers and 88 soldiers had been killed and were buried in the soil of South Africa. The men who had ridden across the Orange Free State and the Transvaal, and had fought at such places as Sunnyside, Koster River and Elands River, had set a high standard for those who were to follow in their footsteps.

In 1906 the Transvaal was granted responsible government. The following year, with the granting of self-government, the Orange River Colony became known once again as the Orange Free State. On 31 May 1910 the Parliaments of the former Boer Republics and those of the British colonies of Cape Colony and Natal agreed on a constitution for a Union of South Africa.

In 1921 a South African War Veterans Association (Queensland) was formed with an initial membership of 700. Each year the veterans held a dinner to commemorate the signing of the peace treaty at Vereeniging. Every ten years a ceremony was held at Government House and loyal greetings from the Association were presented to the Governor for forwarding to the reigning sovereign. The reply from Buckingham Palace would thank the veterans "for their good wishes which were received with much pleasure".[29]

The veterans' final reunion dinner was held in 1969, and with only twenty-nine remaining members, the Association disbanded in 1970. The banners, photographs, medals, and other memorabilia were presented to the Victoria League for Commonwealth Friendship. For many years the ladies of the Victoria League, notably Miss Hutchison (daughter of Colonel K. Hutchison, 2nd Contingent QMI), Jean Hardie, Miss Jean McLeod, and the Patron Mrs Jean McGregor-Lowndes MBE provided support to the surviving veterans. In 1975 six surviving members of the Boer War living in Queensland were honoured by the 2nd/14th Light Horse (QMI) during the Freedom of the City celebration in Brisbane.

Those heroes that shed their blood and lost their lives. You are now lying in the soil of a friendly country. Therefore, rest in peace. There is no difference between the Johnnies and the Mehmets to us where they lie side by side here in this country of ours. You the mothers who sent their sons from far away countries, wipe away your tears. Your sons are now lying in our bosom and are in peace. After having lost their lives on this land they have become our sons as well.

Kemal Ataturk
Quinns Post Cemetery
Gallipoli 1934 [1]

The Great War

Following the Boer War the state defence forces were amalgamated to form the Commonwealth Military Forces. The battalions of the QMI were re-formed as the 13th, 14th and 15th Australian Light Horse Regiments (Queensland Military Infantry). King Edward VII awarded a King's Banner to each unit which served in the South African War, and in 1904 each of the three regiments of the QMI was represented at a Royal Review in Melbourne where the Banners were consecrated and then presented by the Governor-General. (The Regiment's Banners are today displayed in the foyer of the Officer's Mess at Enoggera.) In 1908 each regiment was awarded the Battle Honour "South Africa 1899–1902".

With the introduction of compulsory military training the army was expanded and in 1912 the designation of the units of the QMI were changed to 1st, 2nd, 3rd and 27th Light Horse Regiments (QMI). (During World War I these units were to serve at home while a volunteer expeditionary force was raised for overseas service.) In the early 1900s the Light Horse regiments flourished under the guidance of the army's first Commander-in-Chief, Major-General Sir Edward Hutton. General Hutton was a Boer War veteran and was an advocate of the mounted infantry style of fighting. By 1914 there were twenty-three such regiments throughout Australia.

General Sir Ian Hamilton, the Inspector-General of Overseas Forces spent eleven hot weeks beginning in February 1914, inspecting the Australian Forces. Of all the troops the Light Horse regiments made the strongest impression. In letters to the Governor-General, to the Minister for Defence and to the King's private secretary, Hamilton praised their horsemanship, their daring and their determination. He wrote, "The Light Horse are the pick of the bunch . . . they are real thrusters who would be held by no obstacles of ground, timber or water from getting in at the enemy . . . take them as they are, they are a most formidable body of troops who would shape very rapidly under service conditions."[2] Hamilton's judgment was to be confirmed in the coming years.

In the early days of August 1914, Australia eagerly waited for England to declare war on Germany. This would be the young Commonwealth's chance to prove itself. In the throes of a political deadlock that led to a double dissolution, the Liberal and Labor parties outdid each other in expressions of loyalty to the "Mother Country". By the time war was declared in August, Prime Minister Joseph Cook had promised 20,000 troops, including a Light Horse brigade of 2,226 men.

In the early stages of the war, volunteers could nominate their branch of service, and men flocked to join the Light Horse. In his *History of 2nd Light Horse Regiment Australian Imperial Force, August 1914 to April 1919*, Lieutenant Colonel G.H. Bourne, DSO describes the rush to join:

> The appeals for men met with a ready response. Men representing every walk of life rolled up. Most had had service or were serving in the CMF [Citizen Military Forces], some had had previous active service, some were discharged Imperial soldiers. Some had no training at all. Eleven officers, holding commissions in the CMF, but for whom there were no vacancies in the AIF [Australian Imperial Force] relinquished their commissions temporarily and joined as privates in order to have the honour of being in the First Expeditionary Force. Men were so plentiful that the standard of physical fitness was high, and we had every reason to be proud of the Regiment that later on, sailed from Pinkenbah.[3]

Following a farewell parade through the streets of Brisbane, the Regiment sailed for war on 24 September 1914. Due to reports that the German Pacific Fleet was active in the area, the Regiment was diverted to Melbourne where it was to wait a month before continuing its journey.

During the enforced stay in Melbourne, the Commanding

Officer, Lieutenant Colonel R.M. Stodart, sought and obtained permission from the Prime Minister for emu plumes to be worn as an official part of the dress. A battle honour of the Queensland Mounted Infantry for their work in the shearers' strike, it had been worn with pride by the Queenslanders during the Boer War. The Prime Minister agreed that the emu plume could be adopted as the emblem of the Australian Light Horse and this news was greeted enthusiastically by the troops.

The Regiment re-embarked on 20 October, reaching Albany on 26 October where the fleet collected and took on water and supplies. The 2nd Light Horse (2nd LH) sailed on 1 November in a convoy of forty large transports escorted by British and Japanese warships.

At the time of sailing it was intended that the Australians would be transported to England for training on Salisbury Plain prior to deployment to France; the question of Gallipoli had not yet arisen. Colonel Harry Chauvel, who was the Australian representative at the Dominion Section of the Imperial General Staff in London, was concerned to discover that twenty thousand Australian and New Zealand troops were to go under canvas on Salisbury Plain in the middle of an English winter. Chauvel sought the support of the Australian High Commissioner, Sir George Reid, whose intervention persuaded Kitchener to disembark the force in Egypt. Chauvel's action ensured that the Australians and New Zealanders would train in a mild and healthy climate, virtually rain-free, making the most of the days between disembarking and going to the front.

Egypt

While in transit to England, the Regiment learnt that their destination had been changed to Egypt, and that Turkey had entered the war as an ally of Germany. Little did the men realise the significance of this news. The ship arrived at Alexandria on 9 December, after a voyage of seventy days. The modern equipment of the quays was an eye-opener for the Queenslanders and unloading progressed smoothly. Most of the horses gave no

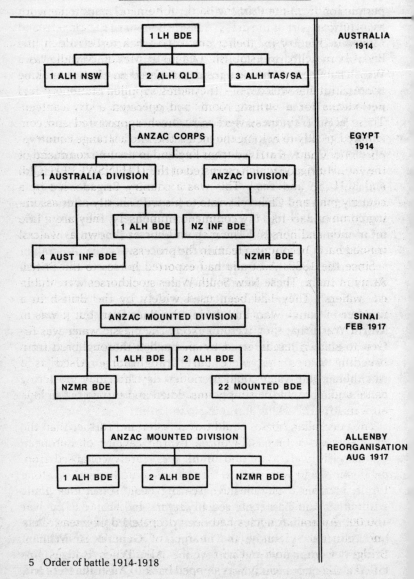

5 Order of battle 1914-1918

trouble – but a few outlaws (notably Billy the Bastard, who was later to become famous at the Battle of Romani) made a name for themselves.

Leading the horses, (being much too weak to be ridden) the 2nd LH marched to establish a camp at Ma'adi. Ma'adi was a British suburb and the local residents spared no effort in making the Australians welcome – the ladies supplied stationery and periodicals for a writing room, and operated a dry canteen. These acts of kindness were very much appreciated and contributed greatly to helping the men settle into a strange country.

Colonel Chauvel arrived from England to assume command of the 1st LH Brigade which consisted of the 1LH (NSW), 2LH (Qld) and 3LH (SA and Tas.). This was a country Brigade, led by a country man and Chauvel wrote to his wife shortly after assuming command: "I like my command immensely; they are a fine lot of men and horses".[4] The men and horses (known as walers) trained hard, becoming a team in the process.

Since the 1830s Australia had exported horses to the British Army in India. These New South Wales stockhorses were dubbed "walers". They had been used widely by the British in a number of native wars and during the Boer War, but it was in World War I that they were to excel. The classic waler was fifteen to sixteen hands, sired by an English thoroughbred from breeding mares that were part draughthorse. Used as a stockhorse, the waler could be ridden day after day mustering cattle and, at night, be simply unsaddled and turned out to look after itself.

The Australian horses could travel faster and further than the heavier coarser breeds favoured by the cavalry of European armies. The walers ate and drank less, rarely collapsed from exhaustion, and were able to carry loads of up to eighteen stone for fifty miles a day through searing heat, sometimes going without a drink for sixty to seventy hours. By the end of the war 160,000 Australian horses had been despatched overseas. Only one returned – Sandy, the charger of General Sir William Bridges, Commander-in-Chief of the AIF. When Bridges was killed at Gallipoli, Sandy was shipped back to Australia to be led, saddle empty, at his funeral. Sandy's head is preserved in the Australian War Memorial in Canberra.

The Light Horse regiments commanded by Chauvel were not cavalry in the European tradition. They were mounted rifles, adept at scouting and screening, for use in operations where speed and flexibility, rather than firepower, were essential. Armed with a rifle and bayonet, they were not trained for that supreme clash of shock action the mounted charge, for which the cavalry was famous. A regiment consisted of about five hundred men formed into three squadrons; a squadron was made of four troops; a troop had ten sections, with four men in each section. The Light Horse were trained to fight on foot, and in action every fourth man was a horseholder whose task was to keep his section's horses under cover ready for the next move. Thus when the Regiment dismounted to shoot or go in with the bayonet, there would rarely be as many as 350 men in action. A whole brigade dismounted would muster little more than a full-strength infantry battalion.

Chauvel trained his regiments hard. Bourne reported, "All day and every day was spent in the desert . . . protection on the move . . . at rest . . . attack and defence . . . night operations, etc, etc . . . till we could have found . . . the landmarks blindfolded."[5] From regimental training the 2nd LH advanced to brigade and divisional exercises.

At this time there was considerable unrest throughout Egypt. The Sultan of Egypt, being pro-Turkish, was deposed by the British government, a Protectorate was established and a new Sultan proclaimed. As a number of Egyptian families were of Turkish descent many supported Germany, and the anti-British movement, backed by German money, armed the desert tribes. To impress the native population the entire British force marched through Cairo.

The Turkish Sultan, who was the worldwide leader of Islam, called for a holy war against the infidel. Accordingly, an attack was ordered with the aim of seizing the Suez Canal, which was vital to British shipping. The success of such an operation would be of immense psychological advantage to the German Kaiser and Turkish Sultan. By January 1915 a Turkish Army Corps had advanced from Beersheba to attack the Canal. The Turks were well-equipped with guns, and carried boats and pontoons. The enemy advance through the Sinai had been spotted from the air and bombed. On reaching the Canal on 3 February 1915, the Turks launched a flotilla of pontoons but these came under heavy fire and only about twenty-five men succeeded in reaching the western bank. Four of the survivors later surrendered in Cairo in response to newspaper advertisements. The enemy occupied some British forward trenches, but the Turks were forced out by counterattacks. By evening, the main Turkish force had retreated into the Sinai after sustaining two thousand casualties and losing seven hundred prisoners. The easy victory overshadowed the Turkish feat of crossing the harsh Sinai dragging steel punts and field guns. This feat roused the admiration of the Light Horse when, some fifteen months later, they were to cross the same desert.

During late February and early March there was increased activity among the infantry units and various headquarters. On the fringe of this excitement the Light Horse waited expectantly; the infantry was about to go to an undisclosed destination but the 2nd LH was not to be part of it. "I am afraid that the Light Horse are rather at a discount in the present war," Chauvel wrote in dejection.[6] The C-in-C Mediterranean Expeditionary Force, General Sir Ian Hamilton was convinced that there was no role for mounted troops among the narrow ridges and steep slopes of the Gallipoli Peninsula, certainly not in the early stages. So in early April, as the Australian and New Zealand Army Corps set off for Lemnos, the frustrated horsemen trained on.

To counter the discontent and frustration a number of inter-brigade manoeuvres were held, together with long mounted treks. While on these operations the troops were given the opportunity to visit the Tombs of Sakhara, and small parties were able to journey to Luxor and see the ancient city of Thebes and the Tombs of the Kings. Novel training included swimming the horses over the Nile at the Barrage on an endless rope, and ferrying them across in native barges.

Gallipoli

Within a week casualties from Gallipoli began to arrive at the hospitals in Egypt. Men sought out friends endeavouring to learn from them something of the objective, the ground, and the nature of the fighting. The casualties had been high, as illustrated by Monash's 4th Brigade which had landed 4,000 strong, but within one week the four battalions could only muster 1,800 men. The Anzac's called for 1,000 volunteers from the mounted regiments to help fill the terrible gaps in the ranks of the exhausted infantry battalions. Dismayed at the prospect of having their commands broken up, Chauvel and Russell (commanding the NZ Mounted Rifle Brigade) volunteered their brigades as dismounted units. Despite the protests of the front-line commanders their proposal was accepted and the 2nd LH avoided dissolution.

The troops handed in their saddlery and mounted equipment and were issued with improvised infantry equipment on 7 May. A day later the dismounted soldiers marched to the Helmiah railway station where they entrained for Alexandria. On 9 May they embarked with 1st LH on the SS *Devanna*. Although enlivened by submarine alarms, the journey passed without incident. At 7.00a.m. on 12 May 1915, the troops were taken aboard destroyers before transferring to the lighters, which were towed to the beach at Anzac Cove by steam pinnaces, landing at about 11.15a.m.

The Regimental Padre, Reverend George Green described the scene:

> No one knew exactly what we were in for. We had vague ideas of the general situation. Nothing official had come to the men in the ranks. Some of us fondly imagined that by this time (it was 16 days after the original landing) the infantry would have advanced some miles inland, that we would be held some distance back, probably awaiting such time as our horses were landed and could be used. What a sudden awakening from such dreams was the first sight of Anzac Cove! There was simply a terrific and continuous rattle of rifle and machine-gun fire and the tops of the ridges seemed ablaze! The Regiment was greeted by a few bursts of shrapnel, wounding four

Shearers' Strike 1891. The officer in the white uniform is Lieutenant Harry Chauvel, who, in the Great War, was to command the Anzac Mounted Division and later the Desert Mounted Corps. (2/14 Light Horse Archives)

The Moreton Mounted Infantry at range practice in the late 1890s. The troops are equipped with Lee Enfield Mark 1 rifles. (Courtesy of Mrs Jensen)

Queensland Mounted Infantrymen circa 1900. (2/14 Light Horse Archives)

This banner, handmade by the ladies of Brisbane, was presented by the Mayoress of Brisbane to the Queensland Mounted Infantry at Meeandah in November 1899, prior to their departure to South Africa. Today the banner hangs in the Regimental Headquarters at Enoggera. (Courtesy of N. Basset)

The *Cornwall* moving away from the wharf into the Brisbane River with the 1st Contingent Queensland Mounted Infantry aboard on their way to the Boer War. *(Queenslander)*

The horse stalls aboard the *Cornwall*. *(Queenslander)*

The memorial cairn at Sunnyside, South Africa. The plaque reads: *"On this spot was fought the engagement of Sunnyside January 1st 1900. This cairn is erected by their comrades in memory of no. 91 Private D.C. McLeod and no. 219 Private V.S. Jones, both of the Queensland Mounted Infantry who were killed in action on that day."* (2/14 Light Horse Archives)

A replica of the Sunnyside memorial. It was unveiled at Enoggera by the Honorary Colonel, Brigadier C.D.F. Wilson on Anzac Day 1989. (2/14 Light Horse Archives)

Private V.S. Jones, one of the first two Australian soldiers to be killed on foreign soil. He died at Sunnyside 1 January 1900. (2/14 Light Horse Archives)

Lieutenant Colonel T.D. Pilcher with officers of the Royal Canadian Regiment and Queensland Mounted Infantry. Colonel Pilcher led the successful attack at Sunnyside 1 January 1900. (Courtesy of R. Thiele)

A group of soldiers with the Union Jack at Elands River August 1900. Despite being surrounded and heavily outnumbered a small force of Australian and Rhodesian troops repelled Boer attacks for ten days until relieved by General Kitchener and his army. (2/14 Light Horse Archives)

A slab of slate from the trench parapets at Elands River used as a headstone in memory of Squadron Sergeant-Major J. Mitchell, Trooper J.D. Duff, Trooper J. Waddell, and Trooper J.E. Walker, who were among those killed during the siege. (Australian War Memorial)

A memorial erected in Anzac Square Brisbane to honour the memory of the Queenslanders who lost their lives in the South African War. (Courtesy of N. Bassett)

Recruits at Chermside camp in 1914. (Courtesy of I. Rayner)

John Markwell farewelling his son "Willie". Major "Willie" Markwell was killed at Beersheba on 31 October 1917 while leading the Regiment in an attack against a Turkish position. (Courtesy of the Markwell family)

The 2nd Light Horse Regiment embarked on the *Star of England*, 22 September 1914. Originally built for the frozen meat trade, the ship was converted to a troop-carrier. Horse stalls were erected on all decks capable of being adequately ventilated. (2/14 Light Horse Archives)

A Light Horse regiment camp near the Suez Canal in 1915. (Australian War Memorial)

A group of 2nd Light Horse Regiment officers in 1915. Second from left is the Commanding Officer, Lieutenant Colonel T. W. Glascow who was later to command the 1st Infantry Division in France. Third from the left is Major T. Logan who was killed at Gallipoli while leading his men in an attack against the Turkish trenches. (Courtesy of J. Bourne)

Cameliers and their camels. The 14th Light Horse Regiment was formed from the Imperial Camel Brigade. Captain Henry Chisholm Mort of Franklyn Vale is on the extreme left. (Courtesy of P. Chisholm Mort)

Ready for action! The horses gave splendid service under appalling conditions, often enduring a lack of food and water for up to forty-eight hours, with an average weight of twenty stone on their tired backs. By the end of the war many of them had travelled in excess of 10,000 miles. (2/14 Light Horse Archives)

The 2nd Light Horse Regiment arriving at Nazlet-El-Abid before starting on the trek south. (Courtesy of the Gerhke family)

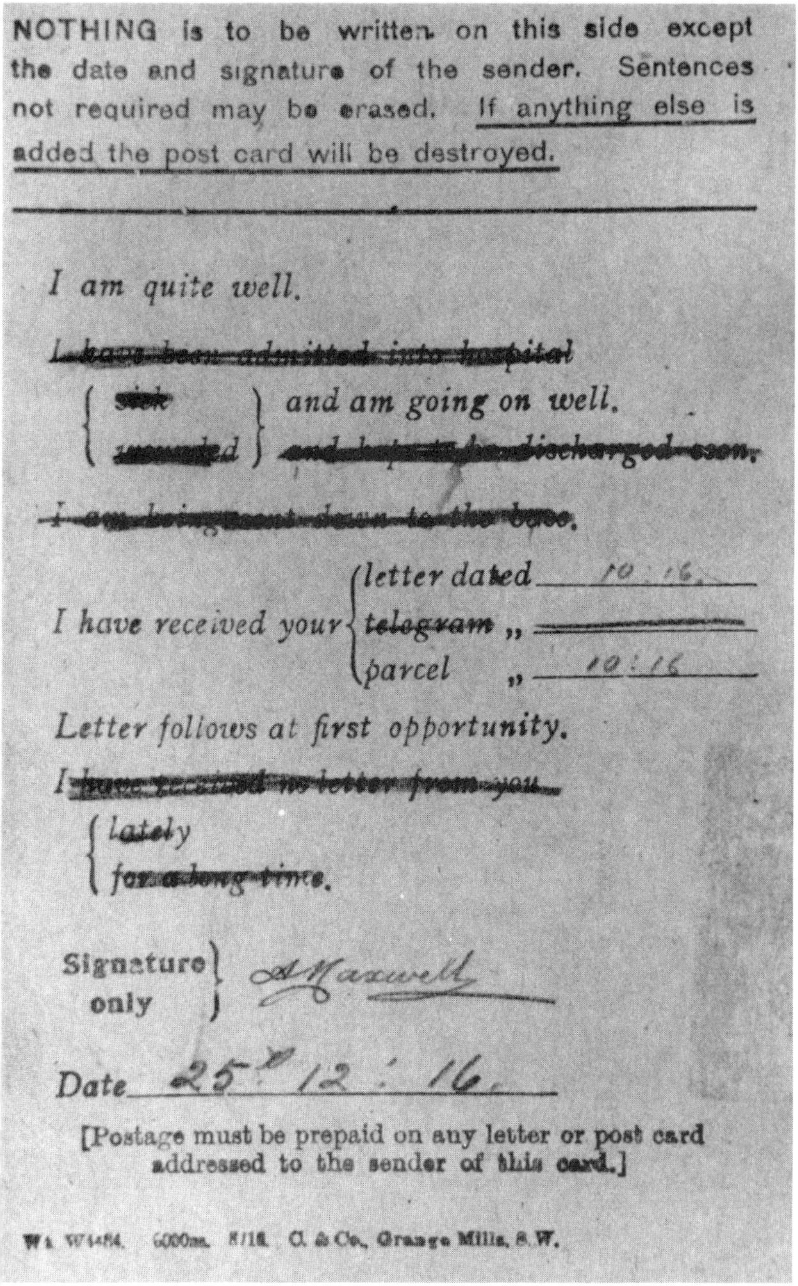

NOTHING is to be written on this side except the date and signature of the sender. Sentences not required may be erased. If anything else is added the post card will be destroyed.

I am quite well.

~~I have been admitted into hospital~~

{ ~~sick~~ } and am going on well.
{ ~~wounded~~ } ~~and hope to be discharged soon.~~

~~I am being sent down to the base.~~

I have received your { letter dated _10.16._
{ telegram ,, _____
{ parcel ,, _10:16_

Letter follows at first opportunity.

~~I have received no letter from you~~
{ lately
{ ~~for a long time.~~

Signature } _A Maxwell_
only

Date _25ᵗʰ 12 : 16._

[Postage must be prepaid on any letter or post card addressed to the sender of this card.]

W̶t̶ W14/84. 6000m. 8/18. C. & Co., Grange Mills, S.W.

A field service postcard sent on Christmas Day 1916 by Anthony Maxwell to his wife. (Courtesy of D. Maxwell)

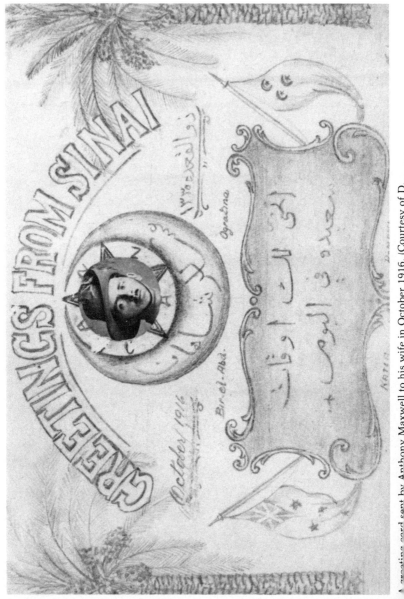

A greeting card sent by Anthony Maxwell to his wife in October 1916. (Courtesy of D.

Horses grazing in a barley field near Gaza. For many horses, after enduring the deserts of Egypt and Sinai, it was the first time in two years that they were able to graze. (2/14 Light Horse Archives)

The left flank outpost of the Ghoraniye Bridgehead on the eastern side of the Jordan River, April 1918. Despite concerted Turkish attacks, the Bridgehead was held throughout the summer. (Australian War Memorial)

Lines of dummy horses were erected in the Jordan Valley as part of a deception plan to convince the enemy that a large concentration of troops remained. Meanwhile the Desert Mounted Corps, under the command of General Harry Chauvel, assembled on the coast. (Australian War Memorial)

Australian Light Horse team creating a dust storm in the Jordan Valley as part of the deception plan. (Australian War Memorial)

men. Trooper Elliot became the first casualty with a slight wound to the head but he was able to remain on duty. There was enough sights on the beach to hold our attention; the evacuation of the wounded; the stacking of the stores; the bathing and shipping activity. We had time to give an exhibition so characteristic of our Regiment, of "justifiable appropriation". In lieu of the leather bandoliers of the Light Horseman we had been given a flimsy drill haversack thrown together in Egypt in a few hours, to do the office of a pack. Quicker than the most efficient QM [Quartermaster] could have managed it, we issued ourselves with regulation infantry equipment from a dump of stuff left by casualties.[7]

That night the Regiment bivouaced in Monash Gully. Before dawn the senior officers were shown around Quinn's Post with a view to taking it over from the 15th Battalion which had held it since 29 April, and was urgently in need of relief. At this stage the foothold at Anzac was roughly a triangle with Quinn's and Pope's at the apex. "It was only 690 acres of land. Only part of the ground was habitable, for much was in direct observation and fire from the enemy . . . Now there must be such things as hospitals, cemeteries, etc, etc and there was over two Division at Anzac!"[8]

Quinn's Post, being at the apex of the Anzac position had the reputation of being the most vulnerable in the whole line of trenches dug on the crest of a semicircular ridge. Enemy trenches were only ten yards away at one point and on higher ground. The main enemy position was twenty yards away, but connected to the 2nd LH position by a series of trenches, these were dug when the enemy position had been captured and held for a short period by the 15th Battalion. The position was commanded by the Chess Board and Dead Man's Ridge on the left and German Officer's Trench on the right, and so could be enfiladed from either flank.

At noon on 13 May the 2nd LH assumed responsibility for Quinn's Post. At the same time the 1st LH took over Pope's on the left flank. Bourne explains the initial few hours:

We had heard little about Bombs, but thought that as they were employed in the Crimean War, and hardly since, they must be more or less obsolete. We were soon enlightened. The enemy bombed us

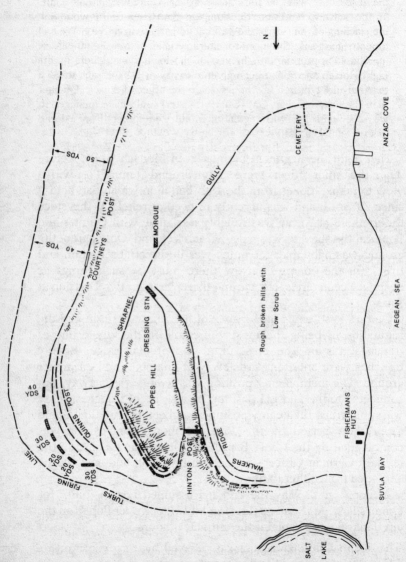

6 Gallipoli, a sketch map of the 2nd Light Horse Regiment's area of action, prepared by H.E. Birch of

well all that afternoon. Our trench was very wide and it was almost impossible for "jacki" to miss dropping his handy cricket ball bombs in. We suffered a number of casualties. Lts Hinton and Boyd and 32 men wounded, till we learned to smother the bombs, or better still throw them back. The need for absolute silence was also impressed upon us. The number of men actually in trenches was reduced to the minimum . . . At dusk the firing line was strengthened but the night passed quietly.[9]

On the afternoon of 14 May, General Birdwood inspected the post and decided, in view of the daily wastage, that the 2nd LH was not strong enough numerically to garrison it. To everyone's surprise the post was handed back to the 15th Battalion who had only had twenty-four hours rest. C Squadron, under Major Graham, was detailed to remain and to assault the enemy position, fill in the trenches between the Turk and Australian positions, damage the enemy positions and return before daylight. There was no artillery or other preparation as the attack was intended to be a surprise.

Several similar sorties had been made by the 15th Battalion and to counter such attacks the Turks had placed machine-guns to sweep the narrow no-man's-land between the opposing trenches. No sooner had the assaulting parties mounted the parapet than they were met by showers of bombs and a tremendous volume of rifle fire. Then the machine-guns on the left and right joined in, with the result that only Lieutenant Ogilvy and three men reached the enemy trenches. Although the enemy had retired, Ogilvy realised the hopelessness of remaining and wisely ordered his men to retire through the old communication trench and to bring in the wounded. Stretcher-bearers dashed out through the hail of lead in gallant attempts to rescue those badly wounded. Major Graham, seeing that the surprise attack had failed, ordered the digging parties to stand fast. He himself scrambled out and ran forward to help the wounded, while doing so this gallant officer was killed. Of the sixty men who made the sortie, twenty-five were killed and twenty-seven wounded. The decimated C Squadron was relieved by a squadron of 3rd LH and then rejoined the Regiment in reserve.

During the following three days the Regiment was employed

digging trenches and preparing a second line of defence. Though only small parties were actually in the firing line, three men were killed and twenty-nine wounded, mainly by snipers. On 19 May the Turks launched a major attack on the whole of the Anzac position. The 2nd LH, less one squadron, was detailed to support the 1st LH on Pope's Hill while B Squadron joined 15th Battalion at Quinn's Post. A total of 42,000 enemy took part in the attack but were successfully repulsed, suffering over 10,000 casualties. The stench that came from no-man's-land was indescribable. "At Quinn's corpses were heaped within three feet of our parapet and swarms of flies divided their attention between the dead bodies and our food."[10]

On 24 May, at the request of the Turks, who sent messengers protected by Red Crescent flags, an unofficial armistice was granted for the purpose of burying the dead. Both sides furnished burying parties while sentries were posted midway across no-man's-land to prevent any observation of each other's trenches. The 2nd LH supplied a party of fifty men, the Reverend Green accompanied the party to read the burial service. Green described the scene:

> Dead in all stages of decomposition were strewn over the ground. In the valley opposite Quinn's and Courtney's the Turkish dead lay as a battalion in open bivouac and one was assured that the accounts respecting the enemy losses on May 19 had not been exaggerated . . . But overall, the stench! It was in one's system for days. My job was the burying. Cotton-wool in my nostrils and occasional nips of rum and water fortified me for the ordeal.[11]

During the next three months the Regiment rotated between the trenches at Pope's Hill, Brigade reserve and providing support to the forward elements. Whether in or out of the front-line the enemy took a daily toll in casualties and, although two drafts of reinforcements had arrived, by the end of July 1915 the Regiment had been reduced to one hundred bayonets, with scarcely a man fully fit. Until then all ranks had made a point of going to the beach when off duty for a swim, as this was the only way to keep moderately clean. However the heavy fatigue work, the long night watches in the trenches, and the tension had so weakened the men that by July few had the strength to walk to the beach.

The Big Offensive

During these last days of July rumours of a fresh offensive by the Anzac troops were rife. The prospect of an advance and bringing the stalemate to an end heartened the troops with the result that the numbers of men reporting sick diminished. All troops not in the front-line were employed in preparing bivouac sites in the rear areas for the arrival of fresh units. The water ration, never very generous, was reduced to one quart per day in an effort to build up a reserve. The great bulk of drinking water was brought from Malta, pumped from ships into barges then to tanks ashore. It was then carried in eight gallon tanks on mules to the unit lines and then by men to the troops in the trenches. Lieutenant H.J. Tiddy recorded in his diary, "I perform my daily wash in a small cup, with this I have to clean my teeth, shave and wash. I manage to clean my teeth with my tea. There is an order to the effect that anybody found washing will liable himself to 12 months imprisonment."[12]

On 4 August the regimental commanders were summoned to General Headquarters (GHQ) to receive orders for the offensive. At this time Lieutenant Colonel Stodart was post commander at Quinn's and so Major Bourne was temporarily in command of the 2nd LH. The plan was that on the night of 6 August a major landing of British troops a few miles north of Suvla Bay would integrate with an Anzac assault on the commanding height of Chunuk Bair. Earlier in the evening, an Australian attack on the Lone Pine plateau would draw attention from this crucial strike. The following dawn the Light Horse would play a key role in a series of feints to aid the attack on Chunuk Bair and, theoretically, be supported by that attack. With hindsight, the whole plan appears to be a framework of unmatched pieces. Enormous responsibility rested on the new troops of Kitchener's army under the elderly General Sir Frederick Stopford who, as Lieutenant of the Tower of London before the war, had never led men in action.

During the evening of 6 August, the lighthorsemen observed the infantry attack on Lone Pine. A sleepless night passed as the men prepared for battle. At dawn the Regiment stood to, awaiting the artillery bombardment on the formidable Turkish positions only fifty yards away. At 4.00a.m. a destroyer sailed inshore and opened fire with a single gun. The shells fell well behind the Turkish front-line. The bombardment stopped at 4.23a.m., seven minutes early, an eternity for the men of the first wave. Turkish reinforcements crammed into the forward trenches. The trenches bristled with bayonets, as a countryman put it: "the way a stubble paddock looks like when you've put sheep across it, and they've turned the earth up a bit, and you see the stubble standing in rows behind the track."[13]

The 3rd LH Brigade, consisting of the 8th, 9th and 10th Regiments, was given the daunting task of charging across the Nek, the causeway ridge at the head of Monash Valley, to attack nine tiers of Turkish trenches on rising ground. The Nek was so narrow that only 150 men could form up abreast, so the 900 riflemen charged in six waves. As they leapt from the trenches they were decimated by heavy machine-gun and rifle fire. Of the first wave only four men reached the enemy trenches and only one of these was to survive. Although cancellation of the attack was proposed, 3rd LH Brigade HQ insisted that further waves be launched. Ultimately, of the 600 men who attempted the charge across the Nek, 372 were killed or wounded. The charge at the Nek was one of the tragedies of Australian military history and tended to overshadow the attacks by the 1st LH Brigade from Pope's Hill and Quinn's Post.

The 1st LH Regiment launched 200 men from Pope's towards four tiers of Turkish trenches. Capturing three lines with bayonets, they clung on desperately for two hours waging a grenade fight with the Turks. When they retired only 46 of the 200 were unwounded.

The men of the 2nd LH Regiment at Quinn's Post faced the worst prospect. Here, no-man's-land was extremely narrow in places, no wider than a road. At the conference at GHQ on 4 August, Major Bourne had been told that in view of the tremendous difficulties attached to the 2nd LH attack, it would not be launched until the German Officer's Trench had been taken, Chunuk Bair had been captured and the Turkish positions on Chess Board and Quinn's were being threatened. These conditions were reasonable as the enemy machine-guns on Chess Board and German Officer's Trench absolutely commanded Quinn's and had been responsible for the annihilation of several previous attacks.

The signal for the attack was to be an explosion of a big mine under the Turkish trenches designed to hurl part of the enemy's front-line into the air and provide the assaulting troops with a foothold. However, at 4.30a.m., the time for the assault, there had been no bombardment, the attack on the German Officer's Trench had failed and there was no evidence of the Turkish positions at Chess Board and Chunuk Bair being attacked. The "big" mine made about as much noise as a "jam tin bomb" and had about as much effect. (The engineers admitted later that they did not get as close to the enemy lines as they had hoped, and that had more TNT been used it would have destroyed 2nd LH trenches.)

Major Bourne ordered the first wave to charge. Major T. Logan and his fifty-five troopers charged into a cross-fire of four machine-guns and massed rifle fire. Before they had run six paces, every man but one had been hit, most in a dozen places. Seeing that the enemy was fully prepared and that further assaults must surely be annihilated, Bourne ordered the second line to stand fast pending further orders. He was concerned that if the whole 200 men detailed for the attack became casualties it would be impossible to repel an enemy counter-attack. Bourne personally reported the situation to Colonel Stodart who obtained Chauvel's approval to defer further attempts. To have pressed the attack would have been a futile waste of lives. The Regiment had already lost Major Logan, Lieutenant Burge and fourteen other ranks killed, and Lieutenant Norris and thirty-six others wounded. Major Logan had served with the 1st Queensland Contingent in the Boer War. A trooper at first, he soon gained his corporal's stripes. He saw action at Kimberley, Paardeberg, Vet River, Zand River, and Diamond Hill. He was wounded in South Africa when a bullet passed through the neck of his horse and on into his own neck. In the same disastrous charge at Quinn's Post, two of Major Logan's brothers, Joseph and John, were wounded, Joseph being hit by no less than five machine-gun bullets. The Regiment had made the sacrifice in forlorn hope of success. Every man knew that victory was impossible but was willing to make the effort to hold the enemy in his trenches, thus assisting the major operations of the British and Anzac divisions.

There was no chance whatsoever of stretcher-bearers working in the narrow no-man's-land. The wounded, if able to move, rolled back over the parapet. It was never possible to recover the dead. Lieutenant Hinton was killed while firing over the parapet, providing covering fire during the withdrawal of the wounded. All day the Regiment continued to provide rifle and bomb support to their front, but it was now obvious that the offensive had failed.

Lieutenant Tiddy's diary recorded the feelings of the men:

> I have had no time to do anything hardly; heavy fighting is still in progress; the dead smell horrible, wounded are still out and an effort is to be made to get them tonight under cover of darkness; we are all tired out. It's now 1.35 am and I'm writing this whilst on the HQ watch. I'm thinking a lot of Home and sick at heart to know so many homes are affected; poor old Tom Logan, however does his wife feel left as she is with 7 kiddies . . . .[14]

Every effort was made to evacuate the wounded as quietly as possible as noted by Tiddy:

> A man is wounded, he is taken at once under cover, stretcher bearers are called for and first aid administered, the man is then taken to the Regimental Medical Officer who fixed him up for his journey to the beach; he is taken to the Beach on a stretcher if unable to walk and handed over to No. 4 Clearing Hospital, anything urgent in the way of an operation is done here. The patient is thoroughly cleaned and made comfortable; the cases such as flesh wounds which only take a week or so to heal are put on board a fleet sweeper which leave here daily at 3 a.m. for Lemnos; more serious cases are put into lighters and towed out to the Hospital ship; as soon as she is full she leaves for Alexandria and then returns. I was on the *Glasconn* a little time ago; she is fitted exactly as a hospital with every comfort and the main

item concerned viz. cleanliness can be obtained which is absolutely impossible on the beach.[15]

For the next two months 2nd LH alternated between the forward trenches, Brigade reserve and work details. On 8 September, Lieutenant Colonel Stodart was evacuated for surgery and Major Glascow assumed command, being promoted to Lieutenant Colonel a few weeks later. Lieutenant Hogue captured the mood of Anzacs, "Pessimism peeped into the trenches. Later, in the solitude of the dug-out, pessimism stayed an unwelcome guest and would not be banished."[16] All the glorious optimism of April, the confidence of May, June and July had gone and the dogged determination of August, September and October was fast petering out.

On 10 November 5 officers and 190 other ranks of the Regiment left for Mudros for a well-earned rest. Those who had been at Gallipoli the longest were selected. Since September, parties of war-worn men had been sent, in rotation for a fortnight's spell, where they could get undisturbed sleep, bathe without being shelled, where rations were supplemented by canteen stores, and there was even the chance of some amusement. The infantry, who had borne the brunt of the fighting, were the first to be spelled out, but then it was the turn of the Light Horse. For the men of the 2nd LH it was an unimaginable release.

On 13 November the Anzac position was inspected by Lord Kitchener and there was considerable speculation as to the reason for this, although the possibility of an evacuation was not entertained by the men. On 26 November orders were issued to vary the daily routine — there was to be complete silence for twenty-four hours — no gun or rifle to be fired, nor bomb thrown. The silence was extended for a further twenty-four hours. The Turks were being conditioned to long lulls in the battle to facilitate an evacuation. The arrangements and orders for the evacuation were worked out in great detail by the staff. It had been estimated that if the enemy received a hint of the withdrawal they would attack with tremendous ferocity and up to thirty percent casualties could be expected.

Ingenious devices were built for firing rifles and bombs after the troops had left the trenches. The withdrawal was carried out over several nights and the mental strain on the last parties to leave was immense. On 18 December at midnight the last of the 2nd LH men (signallers under Sergeant Peterson) left the trenches and wended their way to the beach and embarked for Mudros. By the next morning only 10,000 men remained, one man for every eight yards of the front. They were kept busy simulating normality. The last men were withdrawn early on the morning of 20 December without the loss of one man killed or wounded, ranking the operation as one of the most successful evacuations in modern warfare. This incredible result was achieved by excellent staff work and the intelligent cooperation of all ranks. At dawn of 20 December when the Turks launched a massive attack they found only empty trenches and 10,000 graves.

The Regiment concentrated at East Mudros and embarked on the *Ionian* on 22 December, spending the second Christmas of the war at sea. After landing at Alexandria the men proceeded to their old camp at Heliopolis. Needless to say they were glad to see their horses again and to find many of their comrades, who had been evacuated wounded, now fit and waiting for them.

They were once again mounted soldiers!

Camel Corps

Around the time the 2nd LH was on its way back from Gallipoli, an amazing force was being raised. This was the Imperial Camel Corps which was to operate in the Western Desert around Matruh-Sollum, Dahla and Sinai for the next twelve months.

The use of camels in warfare was not new — they had been used as mounts for Arab archers in the battle of the River Phygius in 190 B.C. and the Romans subsequently used mounted camels both in Egypt and in Palestine.

The British had used camels in the Afghan War (1839), in the Crimean War (1854) and in Abyssinia (1867). Their inexperience in the treatment of camels caused appalling casualties, grimly illustrated by the loss of 70,000 animals in the Afghan War. The Crimean experiment was also a failure as the camels used, accustomed to desert work, could not stand the cold. A Camel Corps, consisting of Egyptian and Sudanese troops, was raised in 1884 for the relief of Gordon. A substantially increased force was employed in the Sudan and in active operations against the Kahlif in 1898–99. But it was not until 1907 that a British Camel Corps was formed in the Sudan.

The Corps was keen to overcome past difficulties. "Camel Corps work was tackled in earnest and the foundation of a corps, its equipment, training and to what uses the corps could be put to, in warfare, was thoroughly tried out."[17] A manual, *Camel Corps Training* was produced in 1913. The Camel Corps conducted long patrols into the desert, made reconnaissance reports on water wells and checked on existing maps – work which was to prove invaluable in World War I. The Camel Corps was the foundation on which that great fighting unit the Imperial Camel Corps (ICC) was built.

The ICC came into being with the return of the troops to Egypt after the evacuation of Gallipoli. In the eyes of the Arabs and the Egyptians the British had lost face by the evacuation and many local inhabitants were ready to flock to the banner of the Turkish victor. Meanwhile, the Nile Valley and the Delta were threatened by bands of Senussi (Arab tribesmen) who, using oases in the Western Desert as bases, would descend into the valleys and disrupt the local population. Experienced Turkish troops, no longer required at Gallipoli, could now be used to take the Suez Canal, thus fulfilling the Kaiser's dream of cutting the British sea route to India. The British had not only to keep both the Senussi and the Turks from invading Egypt, but to settle unrest throughout the country. A force was needed that could fight the Sensussi and patrol the desert oases.

Subsequently under the command of Brigadier C.L. Smith VC, MC, the ICC was expanded to brigade strength. Smith had a wealth of mounted warfare experience gained with the Somali Mounted Infantry and the Sudan Camel Corps, appointed to command the Imperial Camel Brigade (ICB) at the age of 38, he was to remain in command until disbandment of the Brigade in June 1918.

The Brigade was formed into three Anzac battalions and one

British battalion. The strength of the Brigade was approximately 2,800 all ranks. It was able to put in the firing line approximately 1,800 rifles, 36 Lewis guns, 8 Maxim guns, and 6 nine pounders.

The best riding camels are found in the area between Assouan on the Nile, and Port Sudan on the Red Sea. Only male camels were used in the Brigade, it would have been unworkable to have mixed sexes, as in the Middle East mutilation of male animals for sterilisation is forbidden by Islamic Law. The care and use of the camels was drilled into the men at the training camp situated at Abbassia on the outskirts of Cairo. Though much maligned, the camel, if properly treated, can travel for over five days on one drink, which may consist of twenty-four gallons. He can carry five days rations and forage for himself, plus food and water for the rider. The Camel Brigade could therefore travel for nearly a week without requiring a resupply.

The Camel Transport Corps, a different body altogether from the ICB, was supplied with heavy draught camels to carry stores from the base dumps to the front-line. In the advance up the coastal plain in Palestine in November 1917, over thirty thousand camels were used to carry food, water and ammunition to the troops of the eastern force of the army.

The new recruit soon found that reading the *Camel Corps Training* (1913) was one thing, but putting it into practice was another. For instance, when the command "mount" was given, over half of the recruits would be thrown by their camels. The most prevalent problem was biting, but the men soon learnt not to trust their mounts and bites became a rare occurrence after the first month. In addition to mounted training and camel management, all the men were given musketry and semaphore signalling instruction. Following this basic training the troops and their mounts practised the art of long-distance marching. Dropping out to adjust a load meant that to catch up with the rest of the company the unfortunate camelier would be deprived of a well-needed halt, as it was forbidden to "trot" the camel. It normally took one or two days trek before the interruptions by stoppages for adjustments ceased.

While the lighthorsemen were mounted infantry, troops mounted on camels were pure infantry, but with a rapid method of transport. Unlike horses, camels could not be raced close up to the enemy position, nor kept close at hand for a galloping escape if the enemy was found to be too strong. Once the cameliers dismounted they were as committed as ordinary infantry.

The ICB had the distinction of being represented by various detachments in more areas of the Middle East than any other body of troops involved in the campaign. Toward the end of 1915 the ICB was involved in operations against the Senussi in the Western Desert of Egypt. It was during this campaign that Siwa a town situated on an oasis 200 miles south from Sollum near the border between Egypt and Tripoli, was captured. Siwa, like an island in an ocean of desert, was ten day's march over a waterless, uninhabited wasteland.

Australian detachments of the ICB patrolled the oases of Baharia, Dakhla and Kharga, situated to the west of the River Nile. They are a part of the Great Sahara Desert. It was across this country that Cambyses, the Persian conqueror of Egypt, in 525 B.C. sent an army of 50,000 men to try and capture Siwa. The whole force disappeared, without trace, into the desert. The Persians were perhaps overwhelmed by a violent sandstorm, or lost their way and died of hunger and thirst in the desert.

Another detachment of fifty Australians with two machine-guns, made a reconnaissance to Mount Sinai in the south of the Sinai Peninsula, while in July 1918 two companies marched across the Sinai Peninsula to Akaba on the Red Sea. There they joined with T.E. Lawrence and his Arab forces, and trekked north to Amman, from there they made their way back to Beersheba. In addition to all these expeditions, various units of the ICB patrolled the northern portion of the Sinai Peninsula in 1917, while the remainder of the Brigade was to take an active part in Allenby's advance through Palestine and the raid across the Jordan to Amman.

Upper Egypt

Following their evacuation from Gallipoli, the 2nd LH spent two weeks reorganising and refitting. The Regiment was then ordered to Barrage and to Warden where it performed outpost duty for a fortnight.

Early in February 1916 the previous Commanding Officer, Lieutenant Colonel T.W. Glasgow DSO, was appointed to command the 13th Infantry Brigade. He took with him the then Adjutant, Captain Steele and many promising NCOs – quite a loss to the Regiment. Nonetheless their erstwhile mates among the troopers wished them well, for it was known that rapid promotion was possible in the Infantry. Major S.W. Barlow of the 11th LH was appointed to command the 2nd LH.

In concert with other units and the Imperial Camel Corps, the 2nd LH was ordered to the Upper Nile to combat the Senussi Arabs of Western Egypt who, at this time, were threatening the Nile Valley. An expedition, of a purely protective nature, was undertaken by the lighthorsemen to give confidence to the friendly natives of Upper Egypt, and to warn the Arabs that their actions were being monitored carefully.

On 13 February the Regiment moved to El Minya. After leaving one troop at El Minya the 2nd LH marched via Asyut to Sohag. At a number of the larger centres detachments of the Regiment were stationed to conduct patrols in the area. The weather was wretched, with almost daily Khamseens – desert sandstorms.

Warrant Officer D.R. Needham-Walker wrote of the Khamseens, in the *Queensland Digger* of 1 April, 1927:

> During the whole of our stay in the desert, hardly a week passed that we did not experience a sandstorm, sometimes two or three. These storms have not a drop of rain in them, but hot winds which simply lift and drive the sand before them, so that at times it is impossible to see 20 yards ahead. It was not an uncommon thing for the troops on waking in the morning, to find themselves half buried in sand, and the wonder is that there were no cases of smothering in this respect.
>
> These storms would obliterate landmarks, in the shape of hills, as today the troops may have to ride around a sandhill and tomorrow that hill may be any distance away, and the whole appearance of the surrounding country altered, so that when travelling absolute reliance was placed in the maps, compass and stars.

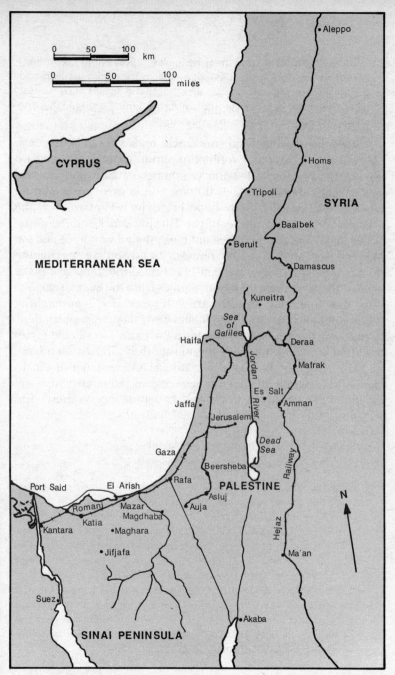

7 The Middle East, the Light Horse's area of operations during the Great War. (Time-Life Books Australia)

Troops ate sand in everything, as it was impossible to keep it out. Their hair was constantly full of it, as well as the eyes, nose, ears and mouth. During summer it became so hot that it would burn the feet through the sole of the boots, and during winter it got so cold that one wondered if it was possible to get warm again.[18]

During the expedition no enemy was encountered by the 2nd LH, but it provided an excellent opportunity to train the reinforcements in mounted work. A number of men died due to sickness and Private Leatch drowned in a canal on 3 March when he was knocked off a narrow bridge by his horse.

Easter 1916 brought news of the Turkish attacks on the posts of the Canal and the Regiment anticipated a move. On 11 and 13 May the Regiment entrained from Asyut and Sohaq respectively, for Kantara. On 18 May the 2nd LH crossed the Canal and marched to Romani, some twenty-two miles from the Suez Canal.

At this time the Turkish Army, numbering over twenty-thousand men and supported by heavy artillery, was within thirty miles of the Suez Canal. It was a far more formidable force than that which eighteen months earlier, had already attempted to seize the Canal. The new British Commander-in-Chief, General Sir Archibald Murray, was committed to an "offensive defence" and prepared to defend Egypt on her border with Palestine.

Romani

Not with bowed head, although a fear
Stain the bright dews of evening there;
But with the love and deathless pride
Their faultless fame has sanctified.

J.B. O'Hara [1]

Sand, Sweat and Horses

Preparation for Battle

To thwart the Turkish advance the 5th British Yeomanry Brigade had been despatched to the vicinity of Romani, with orders to patrol and hold the main wells of the oasis area. At the same time General Sir A.J. Murray took steps to consolidate British possession of the area: a railway line was being constructed from Kantara, on the Canal, towards Katia (five miles east of Romani); and a pipeline, bearing filtered water from the Sweet Water Canal, was snaking out as the army moved forward.

In addition, General Murray wanted to confine the Turkish advance to the coastal route. To this end he ordered the destruction of the wells along the central route. The 3rd LH Brigade conducted a well-executed raid against a Turkish party that was developing new bores at Jifjafa. To support the small raiding force of 90 riflemen a total of 320 officers and men, 175 horses and 261 camels were required, highlighting the massive logistics of conventional desert campaigns, with the Australian waler performing well against the camel. The raid to Jifjafa demonstrated to General Murray that the Light Horse had what he called a genius for this desert life and from now on his policy was to place the Australians in the vanguard of the British force.

A second sortie into the central Sinai by the 3rd LH Brigade destroyed another five million gallons of water. This action ensured that the central route to the Suez Canal was now impassable for the Turks. The only way open was via the oases along the coast. The Turks would have to water at El Arish and then march sixty miles across the coastal sand-dunes to the bountiful group of oases between Katia and Romani before striking at the Canal.

Meanwhile the 5th British Yeomanry Brigade, which had been despatched to Romani, was ill-suited for its mission. The Brigade had not been well-trained and was poorly prepared for isolated warfare in the desert. With the exception of a few officers, all ranks were utter strangers to the desert. To compound the problem, the Brigade was scattered around the area with posts at Romani, Katia, Oghratina, and Hamisah.

On the morning of 23 April a strong Turkish raiding party attacked the British positions, overwhelming the force at Oghratina and Katia. The remainder of the Yeomanry Brigade was ordered to withdraw towards the Canal. The Turks had crossed the desert, defeated and routed a mounted brigade – a success beyond their wildest dreams. After destroying the two posts at once they withdrew with their prisoner, leaving the British wounded at the mercy of the Bedouins. The wounded soldiers were stripped naked, refused water and taunted with cries of, "Finish British". The prisoners were rushed back to be paraded through the streets of Jerusalem as evidence of the invincibility of the Turkish army. Coming so soon after their victory at Gallipoli, the success in the Sinai was of great political and morale value to the enemy.

In response to the Turkish victory the Anzac Mounted Division and British 52nd Division moved without delay to Romani to prevent the Turks from gaining this valuable oasis area. The Anzac Mounted Division had been created in early 1916 from the 1st, 2nd and 3rd LH Brigades and the New Zealand Mounted Rifle Brigade. On 16 March 1916, General Chauvel took command of this impressive fighting force. At a later date the 3rd LH Brigade was withdrawn from the Division which, during the remainder of the campaign, generally fought together. Murray told the War Office, "In any serious mounted work I rely entirely on my Anzac Mounted Division, who are excellent under hard conditions."[2]

It was at Romani, on familiar ground with railhead and water supplies nearby, that Generals Murray, Chauvel, Lawrence, and

Smith wanted to meet the Turks. At this time Chauvel had command of the Mounted Division (less two brigades), Smith was in command of the 52nd Division, while Lawrence was in charge of the overall sector. The commanders inspected the ground and decided that it was an ideal site for defensive purposes. Chauvel's proposal for the defence was adopted: establish a strong camp at Romani; patrol and hold the well-watered area around Katia, by reconnaissance in strength rather than by isolated camps; and, if possible, induce the Turk to accept battle upon the prepared Romani ground. It was believed that the Turk, flushed with his earlier success, would not be satisfied to sit at Katia, but would push on.

Murray and his leaders anticipated in detail, exactly what action the Turks would make. Had they been able to move the Turks like pieces in a game of chess they could not have predicted the enemy moves more accurately. A major difficulty was to arise with the command and control of the battle. Lawrence, who had responsibility for overall command, established his headquarters at Kantara, twenty-three miles from the scene of action and had to rely on an above-ground telephone line for contact with his divisions. Murray realised that Lawrence's HQ was too far away to accurately monitor the fighting. He had "suggested" a move nearer to Romani, or alternatively, placing either Chauvel or Smith in command of both the mounted troops and the infantry at Romani. Unfortunately this was not accepted; subsequently the battle of Romani was not as decisive as it should have been.

Gullet, the official historian, was later to say:

> ... so far as leadership went, General Chauvel won Romani single-handed, despite all the fumbling on the part of the higher staffs .... Never ruffled, even at the most critical moments of the fight, and always forward under fire, the Australian leader confidently imposed on his two Brigades that supreme ordeal of battle ... a slow fighting retirement in the face of overwhelming odds.[3]

Murray and Chauvel were convinced that the Turks, if they advanced, would march south of the infantry posts (dug in front of Romani) and strike at the Romani camps and railway. There was little concern about a frontal attack on Smith's line of infantry posts. Chauvel, with his two Brigadiers and Commanding Officers, examined the country in detail. He selected a defensive outpost line running for three and a half miles south-southeast from Katib Gannit to a point southeast of Hod el Enna. No digging was permitted in order to deceive the enemy into believing this approach was unoccupied and thus entice him into an attack over soft sand. Chauvel aimed at a holding and delaying action and that when pressed, the troops should fall back pivoting from the left on Katib Gannit. It was believed that the Turk attack would become disorganised when they blundered unexpectedly on the Australian line.

The plan was made with great thoroughness, telephone communications were laid out along the outpost line and the second defensive position selected. The idea was that as the Turkish force pressed in towards the camp after the lighthorsemen, its flank would then be exposed to an attack from the south by Chaytor's Mounted Troops based at Duiedar, and the 3rd LH Brigade (under Antill) which was located further south.

Steps were taken to deceive the enemy as to the disposition in the oasis area. To this end a dummy camp for a LH brigade was established at Katia, tents erected and trenches dug to mislead the German airmen. After each day's reconnaissance the forward brigade retired into this camp at nightfall and after a brief halt moved back to Romani under the cover of darkness. The relieving brigade always reached the dummy camp before dawn.

Chauvel continued with extensive reconnaissance patrols, learning a great deal about the ground and enemy locations. At the same time, the mounted troops gained valuable experience of the desert, they became skilled at survival where water was scarce and all provisions had to be carried. All possible marches were made at night, and troops dependent on a single water-bottle would, despite their thirst, rarely drink until the objective had been reached and the return journey started. This was often not until noon on the day following the beginning of their patrol. It was discovered that men who drank their meagre supply of water early, and so were left with an empty bottle, would collapse sooner than men who had not drunk at all but had a full bottle of water in reserve.

Trooper (later Sergeant) Alan J. Campbell of 2nd LH describes the night patrolling:

> Our patrols to the south left after sundown to make contact about midnight with, usually, New Zealand patrols. The New Zealanders, being efficient soldiers, bore the reputation of shooting first, then challenging us on our approach!
>
> Our track upon that southern patrol from dusk until dawn, was as rough a track to ride in the dark as you could ever get. The worst and most common experience was after riding up the fairly smooth windward side of a sandhill, we'd reach the lee side, which was steep and loose, so we would tumble over and over with our horses taking tons of loose sand with us to the bottom. Most unpleasant and disturbing!
>
> Given prismatic compasses, few of us could use them properly, so there were some patrols that never came back . . .[4]

Anzac Ingenuity

During this preparatory phase in the desert, Australian ingenuity solved a number of problems. Disposal of rubbish was difficult due to the fact that sand is weak in bacteriological activity. Great incinerators were constructed from thousands of empty bully beef cans, and these, with careful arrangement to maximise draught, consumed the rubbish from the Romani camps. By contrast the Turkish camps, captured later in the campaign, were littered with rubbish and accompanied by a blue haze of flies. So unhygienic were these captured camps that they could rarely be occupied.

In the oasis areas, water was freely found at a depth of from two to twelve feet, but the quality was poor. Brackish, and made bitter by the palm roots, the water was unpalatable to the men and was refused by the horses until they were extremely thirsty. Lieutenant Colonel Wilson of the 5th LH introduced the "spearpoint" pump, with which he had been familiar in Queensland. This simple device was made of a two and a half inch tube, with a solid point above which was a section of strong wire mesh. The pump entirely changed the practice of watering the horses.

Easily carried on the saddle, in a few minutes it could be unpacked and driven into the sand in a likely spot for water. By the time the others had laid out the canvas water trough, a plentiful supply of water was being pumped out of the sand.

Another difficulty confronting the lighthorsemen was the transportation of the wounded over the soft sand. The sand carts and wheeled ambulances needed to transport the wounded to the rear could make no headway, while the camel cacolets – made of a deckchair device swung on either side of the camel – proved to be an instrument of torture. The Australians devised simple sleighs made of sheets of galvanised iron turned up at the front and sides and drawn by two horses. These proved to be a safe and gentle method of evacuating the wounded. The New Zealanders introduced to each brigade a dental unit capable of treating all simple cases and of repairing/replacing dentures in the field. Able to treat up to 500 patients per month, these units contributed greatly to the fighting effectiveness of the Anzac Division.

Romani

On 18 July, four formations of Turkish troops, about eight thousand men, were observed plodding across the desert, moving steadily towards Romani. Murray and Chauvel were pleased with the appearance of the enemy as it seemed the Turks were being drawn into their carefully prepared trap. Murray wrote to Lawrence, "If the enemy can be induced to come in, it is to our advantage to allow him to do so, and not launch our mobile striking force until he gets committed against the Romani defences."[5] From the beginning of August the enemy line showed signs of increasing activity and aggression, and a firm resistance to the probing of the Australian patrols.

Patrols reported that enemy camelmen had entered Katia just before dawn on 2 August and that they were soon joined by Turkish infantry. Chauvel anticipated that the occupation of Katia was a prelude to an attack on Romani. On the night of 3 August he ordered the 1st LH Brigade to occupy the selected out-

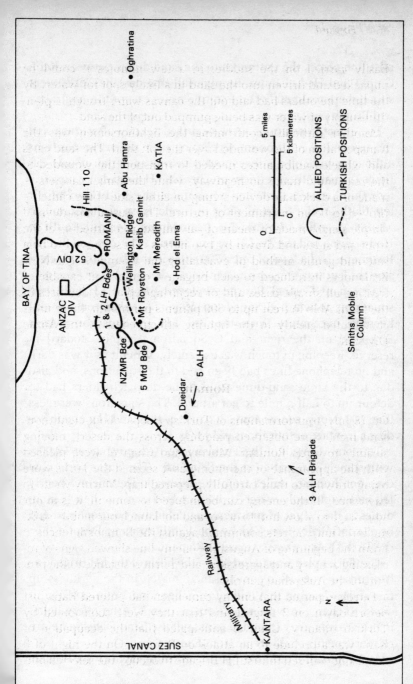

8 Romani – position at 4.00 p.m., 4 August 1916.

post line across the entrances to the sand-dunes between Katib Gannit and Hod el Enna. By nightfall that day the Turks were situated approximately three to five miles east of the Australian and British positions.

The selected positions were occupied shortly after dusk and the troops lost no time in developing their defensive positions although, without sandbags and revetting material, it was difficult to make them effective. The 1st LH Brigade occupied a long thin line over three and a half miles, the 3rd LH, with their left flank on the Romani redoubts, held from Katib Gannit to Mount Meredith, the 2nd LH from Mount Meredith south to Hod el Enna and the 1st LH was in reserve. The 2nd and 3rd LH Regiments had a total of some 500 rifles in the line with which to stop the Turks.

Lieutenant Colonel Bourne deployed the 2nd LH with two squadrons forward: B Squadron (Major Shanahan) on the left, keeping in contact with the 3rd LH, A Squadron (Major Birkbeck) on the right and C Squadron (Captain Stodart) in reserve. Keeping in touch was extremely difficult as it was dark and no telephone lines had been laid to the squadrons, and also due to the large sand-dunes which the despatch riders had to detour up to half a mile to get around. The squadrons were less than 100 strong, so that some 200 men of the 2nd LH were defending a frontage of four thousand yards.

The total strength of the Turkish force was now about 18,000, with some 12,000 rifles in the firing line. The Turks were now over one hundred miles from the railhead. The enemy, poorly fed and wretchedly watered, had trudged over the heavy sand-dunes in the height of summer, dug defensive positions and had been continually harassed by the lighthorsemen. Despite this, the Turkish Army was to excel at Romani.

The Turk's plan was to follow on the heels of the LH Brigade as it returned from its reconnaissance, march in the darkness up the gullies leading to Wellington Ridge, and be in position at dawn to charge down on the Romani camps. After overwhelming the camps, they intended to assault the 52nd Division before reinforcements could arrive from Kantara.

At 9.15p.m. the rearguard of the returning Light Horse brigade passed through the 2nd LH position. At 9.50p.m. the first shots were exchanged between the A Squadron picquet and the leading Turks and within minutes the entire 2nd LH line was being probed from end to end. It became obvious that this was more than enemy patrolling and an all-out assault could be expected. The CO despatched a troop of the reserve squadron to help fill the gap between A and B Squadrons. Meanwhile urgent orders were sent to Major Birkbeck to withdraw his squadron and to take up a position at the rear of the Regiment's left.

About 1.00a.m. 4 August, B Squadron reported that they were being attacked in strength. At the same time the 3rd LH Regiment also notified that they too were under heavy assault. Additional small-arms ammunition was rushed to B Squadron who were ordered to hold on at all costs; should the enemy break through the B Squadron position, the whole of the right of the line, including Regimental Headquarters and A Squadron, would be cut off.

Trooper Campbell described the night:

> At that time of the year the prevailing wind was from the south-west, which gave us on outpost the advantages of compact, fairly firm surface, thus putting the enemy at great disadvantage having to attack us by climbing the dreadful, loose, sandy lee side of the hills. Most of our listening posts were at the foot of those hills and were overwhelmed when the enemy concentrated prior to the attack. We never saw our listening posts again.
>
> The Turks launched their first attack . . . along our whole line. They repeated their attacks several times, to be repulsed by our rifles. Later some enemy officer prisoners stated that our rifle fire was most devastating. The attackers moved in mass formation from the usual light desert bushes, so when they commenced to climb the bright, loose sandhills, the dark forms stood out well, offering us a splendid target to shoot into. As they climbed towards us, not only did they give us a good target on the bright sand, but they moved so slowly in the heavy climbing . . . and their exhaustion helped us repel their attacks. As they retired, their dead and wounded were clearly to be seen.[6]

There were some daring and exciting incidents during the night. Major Shanahan, riding "Billy the Bastard", was checking the listening posts of his squadron when he discovered four of his men without horses and outflanked by the enemy. With three men mounted and a trooper hanging on each stirrup, Billy was forced into a labouring gallop over the soft sand for more than a thousand yards to safety.

The Turks attacked with great strength and determination. At about 1.00a.m. the telephone wire between Lieutenant Colonel Bourne and Brigade Headquarters was broken, the Regiment had no further communication with the Brigade staff until noon. Communication between squadrons was almost impossible and, until dawn, the resistance depended upon the ability of the squadron and troop leaders and the resourcefulness of the men. At 2.00a.m. Bourne threw the remainder of his reserve into the forward line but it was evident that the enemy couldn't be held much longer. At 2.40a.m., no officer being available, Bourne sent an NCO back to select a position to withdraw to; and there was still no word from Major Birkbeck as to whether his squadron had occupied a new position. In the meantime the forward troops were ordered to hold on at all costs.

A major assault on Mount Meredith by some eight thousand Turks forced the 3rd LH back, thus exposing the left flank of the 2nd LH. At the same time, large enemy forces began moving south to cross the front of the 2nd LH in order to outflank the Australians and strike at the railway via Mount Royston.

At 3.20a.m. with the enemy on three sides, the troopers had been forced back to the horses. Bourne now gave the order to withdraw. Trooper Campbell described the action:

> Just before dawn – the Turks finally massed and attacked in their full strength . . . . They were still in massed formation, but with a continuous line unlike their earlier attacks upon limited fronts. This long, unbroken line was able to envelope our single hill position, by advancing along the shallow depressions, unopposed by our posts. From the splendid vantage point upon my horse beside my Major (Shanahan), I could observe the black mass which never faltered. It kept coming, shouting 'Allah! Allah! . . . Finish Australia!' The Turks were shouting all the way, and firing foolishly from their hips.
>
> Finally our men rose to get their horses just over the hill-top, but found the Turks amongst them. The enemy had got around the

hill . . . .. The frantic search for horses, plus wounded and terrified horses rearing and breaking-away from other horses, produced an unforgettable scene. When some of the men got into the saddle, they found a black form holding the reins, so they said, 'Hop up behind!' But the answer would be 'Finish Australia!' which would swiftly bring a rifle down on the black head and so end the argument.

As the enemy pushed our fellows back . . . their exhaustion was obvious, thank goodness! My Major told my mate and I that we should 'Get into the b . . .s with the bayonet!' That really chilled us! That huge black mass against our puny numbers! Eventually we got all our men away. Most were mounted, but some were double-banked, and some were hanging on to stirrup leathers.[7]

Up to this time casualties had been relatively light, although Lieutenant A.S. Righetti of Jondaryan and Lieutenant P.S.R. Woodyatt of Gympie had been killed.

With perfect steadiness the 2nd LH broke away from the Turks. Bourne wrote:

The bullets were making little spurts of flames all round us, owing to the phosphorous in the sand. Here we experienced for the first time the moral effects of turning our backs on the enemy, and the question arose in our minds as we rode, "Can we re-form?" The order "Sections about – Action front" was given as we reached the position, and was splendidly carried out. This high test of discipline gave us renewed confidence in ourselves.[8]

The Regiment was joined by a squadron of the 1st LH which had been sent to provide assistance. At daybreak Bourne observed Major Birkbeck's squadron to the south, laboriously making its way back through the heavy sand. The Regiment withdrew holding each ridge in succession, until the enemy, always working around to the right, outflanked and enfiladed the position and horse lines.

At about 6.00a.m. the 2nd LH Brigade arrived to support the Regiment. By this time the Regiment's line was decidedly mixed with one squadron of the 1st LH and elements of the 3rd LH. The calm and determined efforts of the 2nd and 3rd LH Regiments during the night had virtually won the battle. The Turks had lost six precious hours, the heat was becoming intense, their troops were suffering from lack of water and were exhausted from having been forced to fight for several hours in the heavy sand. Furthermore the Turks were still denied Wellington Ridge which they expected to secure in the coolness of the night, without resistance.

The two LH Brigades fought a desperate battle defending Wellington Ridge. British artillery opened fire and was successful in driving the enemy machine-guns off Mount Meredith. The Turks, with greatly superior numbers, made steady headway. Chauvel, realising how crucial the battle was, did not want to risk the lighthorse brigades in a one-sided hand to hand struggle. He requested a brigade from the 52nd Division to relieve the LH Brigades so that their horses could be watered, and then, in conjunction with the New Zealanders and Yeomanry, conduct an enveloping movement. Such a movement would have had an excellent chance of completely destroying the Turkish force.

The proposal was not agreed to by General Smith of the 52nd Division because he was still under orders from General Lawrence – located many miles away at Kantara. The folly of employing two independent divisions and a remote HQ to defend Romani, was to cost the Australians and British a major success. Chauvel, having no alternative, continued to use his men as infantry while the golden opportunity for cavalry fighting slipped away.

Chauvel was riding around the positions, keeping in contact with his men, ignoring the enemy fire. He had never been a popular leader, but on this day his coolness and courage won the admiration of the men. He was, however, overshadowed by Brigadier Royston who galloped all over the battlefield booming out encouragement. A trooper commented, "He really was quite magnificent and I am sure saved us from what might have occurred at any time in those conditions – an ill-advised and distressing retirement to other positions."[9]

On the afternoon of 4 August the 2nd LH was placed under the command of Royston of the 2nd Brigade. That night the Regiment was committed to another tiring time on outpost duty. The next morning, the Brigade, consisting of the 2nd and 7th LH Regiments and the Wellington Mounted Rifles, formed part of the Anzac Mounted Division's attack. A huge, thin line of Anzacs with fixed bayonets swept the Turks from Wellington Ridge, then charged down its slopes yelling wildly. Over a thousand prisoners were taken, many half-dead with thirst and fatigue. The bulk of the Turkish force made their escape to Katia.

Lawrence at last acted and the orders, for which Chauvel had waited so long, arrived – Chauvel was to take command of all Australian, New Zealand and British mounted troops and pursue the enemy. With the brigades well-scattered it was some hours before this could be organised. The Turks fled, and the 2nd LH, together with the remainder of the mounted troops continued the chase on weary horses. As they advanced from Romani they saw evidence of the enemy's demoralisation: they rode down hundreds of stragglers who made no attempt at resistance, and the desert was littered with weapons, munitions and equipment.

General Murray believed that the enemy was in confusion and that Katia could be easily taken, but General Chauvel was not so optimistic. The Turks had left part of their reserves at Katia and had occupied a natural defensive position protected by a saltpan and swamp. The three Anzac Brigades lined up on the western edge of the saltpan and fixed bayonets for their first cavalry charge of the war. The men forgot how tired they were and revelled in the wild elation of the charge. When the leading horses reached the swamp they became bogged to the knees. As successive waves of riders heaped up behind, the commanders gave the order to dismount. The horses were galloped back to cover while the men laboured through the swamp. Progress was slow, men were exhausted, and before dusk, the whole line was at a standstill. Soon after sunset, Chauvel ordered a general withdrawal.

The Turks made no effort to advance as the lighthorsemen withdrew. The Turks had suffered over 5,000 battle casualties as well as having 49 officers and 3,900 other ranks captured. The Katia engagement had been a rearguard action by the Turks to enable their exhausted troops to escape to Oghratina. Both sides had been reduced to prostration by the desert.

The lighthorsemen were so exhausted that many men slept in the saddle as their horses plodded back to Romani, which was reached about midnight. From the start of the battle on 3 August, the 2nd LH had been in action or at stand-to for fifty-nine hours. Little or no sleep, no rations, only one bottle of water per man, and no water for the horses, it had been a harrowing experience for all concerned.

During the battle the Regiment had lost ten men killed, twenty-two wounded and eight taken prisoner. Four of the prisoners survived captivity but the others, Sergeant Drysdale, Corporals Sommerville and Easton, and Trooper Ward died in the hands of the Turks. On 6 August the Regiment buried the dead and collected material abandoned by the enemy.

Two days later the 2nd LH marched out to attack the enemy's rearguard at Bir-el-Abd, some twenty-two miles from the base camp. A strenuous effort was made to capture the position but the Turks were well-prepared and defeated the Regiment's assault. The attack was abandoned when the Regiment was forced to return for food and water. This was the Regiment's last attempt to molest the Turkish retreat, however it was generally held by the men that had better use been made of the Light Horse's mobility, greater casualties could have been inflicted.

The failures of Katia and Bir-el-Abd soured the brilliant victory at Romani. Murray blamed Lawrence for his remote leadership, while Chauvel believed Murray was at fault for dividing the command at Romani. However, the fact remained that, for a loss of 202 men killed and 928 wounded, the Australians and British had killed 1,250 Turks, wounded 7,000 and taken 4,000 prisoners. Chauvel was quick to acknowledge "a most masterly rear-guard action by the German commander von Kress and his Turks".[10] They had travelled as fast as cavalry, hauling artillery by hand over corduroy tracks of desert brushwood, covered by a rearguard which even the lighthorsemen couldn't outflank. Yet they were retreating and for the rest of the war would be on the defensive.

Following the battle of Romani and the subsequent skirmishes, Murray sent many messages of appreciation to the Anzacs. According to Gullet:

> If the story of the work both before and during the engagement is read only in Murray's own expressions of opinion in the contem-

porary official papers, it is beyond all question that the Anzac Mounted Division fought Romani almost alone. But in the Commander-in-Chief's narrative of the engagement sent to the War Office and subsequently published, the decisive work of the Light Horse is slurred over, and the British Infantry is credited with activities which were not displayed. Still more difficult to understand was the discrepancy between Murray's messages of appreciation to the troops, and his list of awards to officers and men for service covering the period of the Romani fighting. The great majority of these went to troops recruited in the United Kingdom, and an excessive number to the officers of the Staff which had blundered in the conduct of the fight from beginning to end.[11]

Although the Commander-in-Chief later told the Australians that his dispatch covering the operation had been unfair to Chauvel's men, the whole episode caused a great deal of discontent throughout the Light Horse.

There is no doubt that the Battle of Romani was the decisive engagement of the Sinai and Palestine campaign. Until 4 August the Turks had advanced steadily in a carefully planned campaign towards the Canal. The British had been involved in a purely defensive action to protect the Suez. The stand by the 1st and 2nd LH Brigades and the counterstroke by the 3rd, changed completely the fortunes of the Turks. At Romani the Turks lost the offensive never to regain it, and Chauvel's pursuit was the beginning of the British advance. It was the turning point of the campaign. Fighting on the defensive, the British force had first stopped the enemy, and then, after routing him and destroying almost half of his force, had driven him from the prized oasis area back across the waterless desert.

By now the Regiment was suffering from the effects of four months sustained operations in the desert. The health of the men had been greatly impaired by the arduous desert patrolling and night work, it was time for the men and the horses to recuperate. On 30 September, the Regiment moved to Kantara for a six week spell; from here the officers were sent to the Cavalry School, and the men went on leave to Alexandria and Port Said. A welcome break for all.

El Arish

From Bir-el-Abd the Turks withdrew their main force across fifty miles of waterless and inhospitable desert to El Arish, leaving behind a strong outpost at Mazar, twenty-four miles east of Abd. In addition to this there were Turkish garrisons at Nekhl and Maghara in the great barren range of the Central Sinai, to the east of Hassana and Kossima. Consequently the enemy was in a position to menace the right flank of any advancing force.

These garrisons were weak and, by the middle of August, General Murray had assembled a sufficiently strong force for a march on El Arish. By this time both Murray and Lawrence realised the dangers of desert warfare without adequate water. The railway line and water pipeline would have to be extended all the way to El Arish. The rate of advance of the British forces was to be governed by the speed with which the railway and pipeline could be pushed forward. In spite of a vast supply of Egyptian labour, it took almost a year to build the railway line the 110 miles from Kantara to El Arish.

The advance across the desert was achieved by the mounted troops conducting a reconnaissance in front of and on the flank of the creeping railhead, while the infantry marched from base to base along wire-netting tracks laid on the sand. This valuable improvisation had come from the Australian bushmen who had used a similar technique at home to cross dry sandy creek beds during the dry season.

Throughout 1916 the internal affairs of Egypt made a heavy demand on the Commander-in-Chief. The Egyptian unrest, coupled with the demands of a now widespread command were such that General Murray could no longer exercise effective command from Ismailia. With War Office approval he withdrew his headquarters to the Savoy Hotel in Cairo, over 140 miles behind the advanced force at Romani. Major-General Sir C.M. Dobell was promoted to Lieutenant-General and given command of all troops east of the Suez Canal. This very inexperienced General now commanded a force of infantry and cavalry substantially larger and more difficult to handle than an infantry army corps.

In other moves General Lawrence proceeded to the United Kingdom, he was replaced by the strong, sympathetic and outstanding tactician, Major-General Sir P.W. Chetwode. Like most British Generals, Chetwode took some time to gain a correct appreciation of the Australians. Soon after his arrival in Sinai he complained to General Chauvel that the lighthorsemen did not salute, and even laughed at his mounted orderlies whose smart dress and stiff horsemanship contrasted with the casual and desert-worn Australians.

On 15 November the 2nd LH, which had been training and recuperating on the Suez Canal, rejoined the Anzac Mounted Division. By now the Egyptian Labour Corps was building the railway at the rate of one mile a day, and the water pipeline kept pace with it. This water main was able to water 100,000 men and 30,000 horses. From the pipehead, the water was taken forward in twelve gallon cans, called fantassies, these were carried in pairs on camels. The railway and water pipeline were two of the principle factors in the success of the campaign.

The enemy was known to be in strength at Mazar, about forty-four miles east of Romani. An assault by the 2nd and 3rd LH Brigades in early September had been unsuccessful. The attack was to have been supported by Australian companies of the ICB, commanded by Captain G.F. Langley (later Commanding Officer of 14th LH), however, due to the difficult country, the camels did not reach the position until after the Light Horse had withdrawn.

During the withdrawal an ambitious scheme had been devised to water the two Brigades in the desert. Over fourteen thousand gallons of water had been dumped, but no organisation had been established to distribute it when the horses arrived. When the Regiment arrived there was a mad scramble at the troughs, with the result that many horses were not watered and were forced to go thirty hours without drinking.

Patrols by the Light Horse to the rear of Mazar, in search of water, caused the Turks to become nervous. They soon abandoned the position and moved back to El Arish.

As soon as Chetwode learned of this retreat, he ordered an advance on El Arish. Orders were issued to the Anzac Mounted

Division and the new ICB to march during the night and envelop the enemy before dawn. The 1st LH Brigade was the advance guard and secured a position to the east of the village; the ICB was to the south; the New Zealanders to the southwest; the 3rd LH Brigade went to Masaid; while the infantry moved along a road direct to El Arish. As day dawned the village was surrounded, but the Turks had fled to Magdhaba some twenty-seven miles away.

Although the operation passed without incident it marked an important phase of the whole campaign. For the first time the mounted men were not hampered by the sands of the desert – they had gained a foothold on the very fringe of southern Palestine. "That night" wrote Brigadier Cox of 1st LH Brigade, "will always seem to me the most wonderful of the whole campaign. The hard going for the horses seemed almost miraculous after the months of sand; and as the shoes of the horses struck fire in the stones in the bed of the wadi, the men laughed with delight. Sinai was behind them."[12]

Gullet wrote of El Arish:

> At the end of 1916, El Arish had been in Turkish possession for more than two years, and the inhabitants of the old mud-built village, which is the largest centre of population upon the Sinai Peninsula, had probably begun to look upon the change in government as permanent . . . but whatever their true feelings in regard to the Turkish evacuation, the dramatic appearance of the Anzac horsemen encircling the town at dawn on the morning of December 21 caused the helpless, time-serving Arabs to greet the Australians with an excited demonstration of delight. As Captain Hudson, Staff Captain of the 1st Brigade, rode into the dusty village, the natives hailed his party as deliverers . . .![13]

Magdhaba

Following their withdrawal from Mazar the Turks continued eastwards. On the morning of 21 December airmen reported that the enemy had commenced work on a series of defensive positions around Magdhaba. The following day, a force of ten

Australian airmen raided the town dropping 120 bombs and receiving heavy rifle and machine-gun fire in return. Chetwode had prepared for a simultaneous advance towards both Magdhaba and Rafah, but on receiving the airmen's report decided to use all of his mounted force against Magdhaba. The 52nd Division, who were in the process of marching into El Arish, were ordered to hold the base in the absence of the lighthorsemen.

The advance guard marched fast and it was interesting that the horses, moving on firm ground for the first time since they had arrived in Egypt, often over-reached and stumbled. Shortly before 4.00a.m. the Turkish camp fires were spotted and the column closed up.

Once the brigades were in the assembly area the horses were fed and the men ate breakfast. Meanwhile Chauvel, accompanied by his commanders and staff, conducted a reconnaissance of the Turkish position. At about 6.30a.m. the airmen arrived and commenced bombing the Turks. The enemy, firing at the pilots, disclosed their positions to Chauvel. Shortly after 8.00a.m. Chauvel received the first aeroplane report detailing the enemy's defences and advising that no enemy reinforcements were in the vicinity. Within thirty minutes the brigades were on the move into position for the attack. The 2nd LH, having been detailed as the Brigade reserve, was not engaged as a unit but subsequently assigned a number of tasks. C Squadron was assigned as the advance guard while B Squadron was to play a vital part in the latter stages of the attack.

The enemy was brilliantly sited in five redoubts on a forbidding plain protecting the only water for many miles. Chauvel and Chetwode were aware that the position must be taken or the entire force would go thirsty for at least twenty-four hours. The 3rd LH Brigade swung north and linked with the New Zealanders. The 1st Brigade and Camel Corps struck from the northwest and west. The first moves drew heavy fire from the Turkish redoubts. After the rapid mounted advance, Chauvel's men were making slow progress on foot, across the open country against heavy fire. Chauvel wrote of the victory at Romani, "It was the empty Turkish water-bottle that won the battle."[14] Now it was the lighthorsemen whose water-bottles were empty. They (and their horses) were now twenty miles from water unless the wells at Magdhaba could be captured.

Early afternoon, after little progress had been made, Chauvel decided to break off the attack and withdraw, but General Cox of the 1st Brigade was about to launch a bayonet charge when the signal arrived. Demonstrating his well-earned nickname "Fighting Charlie", he barked, "Take that damned thing away and let me see it for the first time in half an hour."[15] The 1st Brigade and the Camel Brigade literally raced towards a redoubt and the first Turks surrendered.

Suddenly Chauvel had a victory on his hands and the men poured in on Magdhaba from every direction. A vital role was played by Major Markwell who gathered some men of the 3rd LH and B Squadron 2nd LH and stormed the chief remaining enemy redoubt. Meanwhile, Major Birkbeck skillfully led his troops to the rear of the position and threatened a mounted attack on the enemy's only line of retreat. Lieutenant Given of the 2nd LH, raced with his troop into Magdhaba itself just ahead of the 3rd Brigade.

The entire enemy force of about 2,500 was either captured or killed, at a loss of only 22 men killed and 124 wounded. The wells, however, could not cope with the entire force. The 1st Regiment was detailed to clear the battlefield and remain at Magdhaba. Small camp fires were lit beside each group of wounded to warm them and to guide the ambulancemen.

The wounded were loaded in camel cacolets and a long column of 150 camels set off towards El Arish carrying the jolted, groaning men. At El Arish the men were lifted down and rested. They were still thirty miles from the railhead and the higher staff had not made any suitable arrangement for their evacuation. After the wounded had been packed into sand-carts for the long journey to the railway, orders were received to evacuate them by sea. However, due to strong, bitterly cold winds and heavy seas, evacuation by ship was later deemed impossible. Thus, some five days after being wounded and exposed to bitter winds and freezing nights, the wounded were once again loaded onto sand-carts and the column set off for the railway as originally planned. Consequently, from seven to nine days elapsed between the battle and the wounded arriving at a hospital. Despite having conducted hundreds of campaigns over the years the British staff still failed to provide adequate support to the men wounded in battle.

The 2nd Regiment was detailed to escort the prisoners to El Arish, this entailed another night march. As the men had virtually had no sleep for the previous eighty-four hours the journey was extremely difficult: men went to sleep on their horses, – some falling off; others hallucinated, chasing imaginary figures across the desert; many men told of riding through exotic towns and of seeing unearthly creatures. The Regiment arrived back in El Arish on 24 December.

The prisoners were handed over to the 52nd Division and, in order to feed them, the men of the Scottish Division had to go short of rations for a day or two. The "Scotties" were amusingly indignant, and repeatedly told the men of the 2nd LH "they had taken too . . . many prisoners, and should have used the . . . bayonet more".[16]

Christmas was not a cheerful day. The men spent the time, when duties permitted, trying to make up for lost sleep. Although everyone attempted something special for Christmas dinner, with little other than rations this was difficult. But to top it off, the rain was so heavy that Christmas dinner had to be consumed by the men standing up, holding the food and mugs at arms length to avoid the torrent pouring off their hats.

Rafah

Following the success at Magdhaba, the British force spent two weeks recuperating and refitting before pursuing the Turks towards Palestine. El Arish, with a stretch of some thirty miles of heavy, almost waterless sand between it and the railhead, was neither safe nor comfortable. The Turks, on the other hand, had a well-watered route from their position to El Arish. Chetwode was not happy with the situation and, as the brigades of the 52nd Division arrived, he set them to work fortifying El Arish.

It was believed that the enemy still occupied strong defensive positions near Rafa and Magruntein, some twenty-five miles from El Arish. On 30 December the 1st Light Horse Brigade conducted a reconnaissance to determine the Turk's location and strength. Lieutenant Colonel Bourne reported, "We arrived at Sheikh Zowaiid before dusk, and just in time to allow Brigadier-General Cox and his escort to ride towards Magruntein and carry out the reconnaissance. The main enemy position was found to be in this latter place and about two and a half miles southwest of Rafah. The garrison was estimated at about one thousand."[17]

On 8 January 1917 the mounted troops of the Desert Column (the Anzac Mounted Division and Imperial Mounted Division), with the addition of the 1st, 2nd and 3rd Battalions of the ICB, set out to raid the Turks. Chetwode was in overall command, with Chauvel in command of the Anzac Mounted Division and the ICB.

The advance march was an all-night affair. During the night the men came upon grass and let their horses graze, for many it was the first time since leaving Australia. At dawn they passed the Palestine border marker and viewed a countryside covered in grass, clover and barley.

Rafah was a daunting sight. It was an ant's nest of fortifications, dominated by a hill (Hill 255) of tiered fortifications. As dawn broke, the Turks discovered long columns of horses and camels moving rapidly to completely encircle them. Though the enemy were surprised, both Chauvel and Chetwode were to later admit that the strength and bareness of the position disheartened them so much that they did not believe victory was possible.

The 1st Brigade attacked from the north, and was first in action. The Camel Brigade attacked from the east and the Yeomanry moved in from the south. The 3rd Brigade (less the 8th Regiment) was sent in between the 1st Brigade and the Camels, while the 8th Regiment was sent to keep the enemy force under observation and to intercept any reinforcements sent in by the Turks. The 1st and 2nd Regiments went in first, with the 3rd Regiment behind held as reserve.

The 2nd LH reached a road that was sunken in parts. This provided scant but welcome cover in an area lacking any other shelter. The Turkish fire was heavy and the New Zealand troops, forced to travel further than anticipated, took a long time to get into position. The Camel Corps was under pressure and the Yeomanry had been forced to withdraw. A Turkish force was reported to be advancing from the north, and a regiment was sent to intercept it.

The whole situation seemed critical, and General Chetwode believing his forces defeated, ordered a withdrawal. This order, however, was not obeyed as the New Zealanders, failing to receive it pressed on. Presently a few Turks, afraid of the bayonet in the dusk, surrendered. This was the signal for the 2nd LH to push on, and when the message was passed, first to the CO then to the Brigade Major and finally to Divisional Headquarters, that the Light Horse was going ahead with every hope of success, the order to withdraw was cancelled. A general advance was ordered instead, and this met with complete success.

The Regiment had the joy of conveying to HQ the news that Rafah had fallen (theirs being the only regimental telephone in the division that remained functional). In fact Colonel Bourne acknowledged in his records the excellent work of Lieutenant Letch, the Regiment's Signalling Officer, and of Signaller Mercer, in keeping the telephone repaired under fire during the battle.

The total British casualties were 71 killed and 415 wounded. The enemy lost 200 killed, 168 wounded and 1,434 taken prisoner. The captured weapons included four mountain guns and a number of machine-guns.

After this battle the troopers had the opportunity of seeing the Bedouins at work, swarming over the battlefield and stripping dead and wounded of everything of value. Uniforms and boots were torn from the dead, and from any wounded Turks who remained on the field. Almost all of the Australian and British wounded had been taken from the field during the battle, but, despite giving all possible aid to the wounded Turks after the battle ended, some of them were still awaiting help when the Bedouins arrived. These raiders even opened graves to steal whatever may have been interred with the bodies.

Gullett believed that throughout the entire desert campaign, the British policy pandered to the roaming Arabs of western Palestine:

> But the Foreign Office, entirely ignorant of the quality of these people, insisted that the army leaders should treat them as respectable practising Moslems, kin to the Arabs of the Hejaz and of the same faith as the Moslems of the Indian Empire; and instructions were given that special care must be taken not to offend their susceptibilities. The Bedouins, who were almost entirely without either moral or religious principles of any kind whatever, readily took advantage of the situation. For more than two and a half years they continued to engage with impunity in thieving and more serious crimes against the British forces, and to bring false charges against the men to the sympathetic ears of the British Staff Officers.[18]

Gullett was adamant that firm punishment for gross offences at the outset would have saved "infinite trouble later on, and the loss of many good Australian and British lives by murder".[19]

After the Rafah engagement, the Sinai was clear. Chauvel's horsemen had sent the Turks from the desert, and had secured a foothold on the soil of the plains. The Turks fell back to the Gaza-Beersheba line, where they secured a strong defensive position based on an adequate supply of water. For some weeks they left a 4,000 strong garrison at the carefully prepared We'Li Sheikh Nuran, some short distance west of Shellal. However the results of the battles at Magdhaba and Rafah had caused them to have a healthy respect for the Anzac Brigades and the Camels, so early in March 1917 this force was also withdrawn.

Bourne's forward leadership was cool and admirable, and the work of his 2nd Regiment of Queenslanders, on whom fell the brunt of the resistance, was never excelled in the career of the light horse.

H.S. Gullet [1]

Their Finest Hour

Defeat at Gaza

As the ponderous yet remarkable arteries of the campaign, the pipeline and the railway, edged into Palestine, the force prepared for the next strike at the Turks. A new second mounted division, the Imperial Mounted Division was created, incorporating the 4th, 11th and 12th LH Regiments. The remaining brigades, together with the addition of a Yeomanry Brigade, stayed under Chauvel's command in the Anzac Mounted Division.

The 2nd LH was withdrawn from the front and employed in communication protection duties – including protection of the supply dumps, wadis and pipeline. On a number of occasions German airmen would land alongside the pipeline, blow it up and then fly back to their own lines.

During this time the first battle of Gaza was fought (an event the Regiment took no part in). The Turks, under command of von Kress, had drawn back to a scattered twenty-five mile line across southern Palestine, between the fortress Gaza and the frontier town of Beersheba. Gaza was all but impregnable, flanked by commanding ridges which had been expertly fortified.

This attack on Gaza, under the command of General Dobell, involved the mounted troops encircling the town to prevent retreat and to cut off reinforcements, while the infantry made a direct assault. The failure of this sound plan can be attributed to a divided command structure and lamentable staff work.

The British forces, by nightfall, succeeded in reaching the crest of the dominating peak of Ali Muntar. This was despite poor command, a lack of coordination between the attacking forces, and the failure of the artillery to neutralise the enemy. The Turkish commander signalled von Kress that he could not hold without reinforcements, while the townspeople began to prepare a feast to welcome the British conquerors. Dobell, however, perturbed by the arrival of 6,000 Turkish reinforcements and, regardless of the fact that these were being successfully contained by the 3rd LH Brigade, ordered a withdrawal! Despite protestations from all levels of command (Chauvel declared, "But we have Gaza") the order from Dobell was confirmed. Von Kress, encouraged by the unexpected victory, rushed in reinforcements and extended the defences. The Turks could not believe their good fortune. Bourne described it as a "victory turned into defeat by want of nerve".[2]

On 8 April the Regiment moved to Khan Yunis, the point of assembly for the second battle of Gaza. There, the Regiment drew picks and shovels and practised with the recently issued Hotchkiss Automatic Rifles. The picks and shovels would be needed now that they were moving into harder country. In addition there was gas drill with the newly issued gas masks; it was anticipated that the Turks would use gas in the coming battles.

Early in April the Turks had a force of twenty to twenty-five thousand riflemen on a sixteen mile front that extended from the sea in the west, through Gaza and then southeast to Beersheba. From this line the Turks were in an ideal position to observe any movement from the south and southwest. The position was protected across the front of Gaza by a maze of cactus hedges on a slope known as Samson's Ridge.

The more the lighthorsemen studied the dominating and carefully spaced redoubts, with their overlapping fields of fire to sweep the bare slopes, the more it reminded them of Gallipoli. The plan was for the British infantry to launch a massive attack on a five mile front while the Desert Column swung eastwards to prevent reinforcements being sent to Gaza. Although Dobell was stretching his resources he had some surprises: over two thousand gas-shells (to be used for the first time in this theatre), the "war-winning" tank and the new portable Hotchkiss rifle.

The British infantry moved forward and dug in as the Light Horse probed forward in a scouting role to draw fire, locating the enemy positions. The 2nd LH, together with the 3rd LH, was held in reserve ready for use as an early break-through was expected.

The battle began at 5.30a.m. on 19 April, with a bombardment from three warships and the massed artillery. The Turks were well-prepared and the only success of the shelling was a hit on the Great Mosque, which was being used as an ammunition bunker. The gas-shells proved useless in the high temperature and coastal winds.

That afternoon the Regiment was ordered forward to relieve the 1st LH and be ready for a bayonet charge at dusk. Bourne reports, "We trotted up under shell fire, sent our horses back and commenced to takeover just as the Turks advanced against our position. Major Chambers was wounded while laying out trenches for his Squadron to dig. He was dressed by Captain Machin (RMO) but died at the Field Hospital. In him we lost a gallant and deservedly popular officer."[3]

By now the Turks were shelling the men and their cavalry was threatening the rear and the horses. But the new Hotchkiss guns performed admirably and the Regiment was able to repulse the attack. Though the 2nd LH held its position the situation was deteriorating on their left flank. Then, without consultation, a brigade staff officer ordered the Regiment's horses forward in preparation for a withdrawal. As the horses were led, in broad daylight, to the firing line they attracted heavy shell fire and suffered many casualties. The CO ordered the horses rearward and awaited the command for withdrawal. The unwelcome order was received a little later and the horses were moved forward under the cover of darkness, but this time under Regimental arrangement. The position was evacuated without difficulty, having been held for seven hours against stiff Turkish action. The following day, during a move in conjunction with the remainder of the Brigade, the column was heavily bombed by three enemy aircraft. The Regiment lost two men killed, seven wounded and forty-one horses killed during the air attack.

While the 2nd LH was playing its part in the attack, the 1st Battalion of the ICB, commanded by Lieutenant Colonel G.F. Langley, was employed in an attack against a Turkish redoubt. As the men of the Camel Corps advanced they found little cover, except for small wadis and an occasional hollow. A British tank, tasked to support the cameliers, took up the lead and the troops instinctively swung in behind it. Almost immediately it became the target for every Turkish gun in the area and the following troops suffered heavy casualties. When the attack had begun the two leading companies numbered about two hundred men, but the shelling attracted by the tank had reduced their numbers to about one hundred.

The tank headed straight for an outpost some one hundred yards in front of the main Turkish trenches and, in spite of being constantly shelled, managed to crush the barbwire and drive up onto the highest point on the redoubt. Then, hit many times, it exploded into flames. Taking advantage of the brave action by the tank crew, the Cameliers rose and charged forward with their bayonets at the ready. The position was occupied by about six hundred Turks together with German and Austrian officers, but the Australian and British troops fought with superhuman strength against seemingly insurmountable odds. Many of the enemy were killed while the rest panicked and fled. Captain Campbell of No 2 Company ordered his six Lewis guns into position and many Turks were mowed down before they reached their trenches.

The remnants of the Camel companies and British infantry held the knoll for over two hours, but suffered heavy casualties. Campbell ordered the few remaining men to withdraw until he and Lieutenant Aylwin (later 14 LH) were the only two Australians left. Gullett reported, "Campbell and Aylwin had enlisted in Toowoomba together at the beginning of the war. As the Turks drew very close Aylwin made his dash, followed by Campbell, under very heavy fire. Aylwin was hit as he ran, but Campbell's luck still stood, and he was one of only five men out of the 102 who made up the company in the morning, who did not become casualties."[4]

By nightfall on 19 April the Light Horse had abandoned the

few positions they had captured. The Anzac Division had driven back a few demonstrations by large forces of Turkish Cavalry, but they had gained little ground. The bombardment had failed, the gas had failed, the tank had failed, and the tactics had failed. Despite superhuman effort by the men they had been given little chance of success. British casualties were over 6,000 but these were disputed and figures of 10,000, 15,000 and even 18,000 were quoted and believed by the officers and men. Without doubt it had been the most costly day of the campaign. The lighthorsemen had lost faith in the British High Command and as the Australians joined the infantry in the withdrawal words like "murder" and "death-traps" were expressed by the men. Gaza had never been seriously threatened and Turkish morale now soared; whatever ideas their command had entertained of withdrawing were discarded.

There were repercussions in the British command structure. Dobell was relieved of his command, with Chetwode succeeding him as the leader of Eastern Force. Chauvel succeeded Chetwode as the commander of Desert Column, thus receiving the well-earned distinction of being the first Australian soldier to attain the rank of Lieutenant-General. Chaytor of the New Zealand Mounted Rifles Brigade was appointed to head the Anzac Mounted Division.

Shortly after Murray implemented these changes he himself was relieved of command and replaced by General Sir Edmund Allenby. As a commander, General Murray had achieved much, despite his failure to grip and direct Eastern Force. Starting with a totally inadequate force he had shattered the Turkish offensive against Egypt and cleared Sinai of the enemy. Always short of men, munitions and engineering supplies, he had been responsible for laying the railway and pipeline across the desert to the watered country of southern Palestine. The magnitude of his work in conquering Sinai can be gauged by the fact that at the end of February 1917 he had been responsible for the laying of 338 miles of railway, 300 miles of water pipeline, 200 miles of metalled road, 86 miles of wire and brushwood road, while 960,000 tons of stone had been won from distant quarries. "He was a pioneer . . . and he reaped the harvest which is so often the pioneer's bitter reward. In warfare there is occasionally a feather-edge between brilliant success and disastrous failure."[5]

During the remainder of April, May and most of June 1917 the Regiment was employed on outpost duties, patrols and improving the defences. On 18 May a patrol of two troops succeeded in penetrating right up to the Turkish line. They drew heavy fire and one man was killed. Corporal Geddes made a gallant effort to bring him in, but the fire was too heavy and he was unable to move him. Due to continual operations the Regiment was now very weak numerically and even with the arrival of Lieutenant Brett and thirty-eight other ranks the unit remained far below strength.

On 18 June the Regiment, together with the remainder of the Brigade, moved to Marakeb on the coast. For ten days the men had a holiday with plenty of rest (no broken nights), swimming and sports. It was a place of many happy memories for the men of the 2nd LH. At night amateur theatre and boxing tournaments entertained the troops, taking their minds off the war. During the rest period the men were deloused, using railway carriages as fumigation chambers. Body lice were a continual problem, and they were not only an irritation but spread typhus.

During the next four months the Regiment was actively involved in preparation for the third battle of Gaza, or the charge at Beersheba as it became known. This involved patrols, providing protection parties to reconnaissance groups, and the laying of underground telegraph cables. It was during a patrol to capture Turkish prisoners that a member of the Regiment was awarded the Military Medal.

The Regiment had moved out into no-man's-land, which was approximately ten miles wide, during the night and had formed a line three miles long in order to sweep back towards their base camp. As dawn broke the men discovered two regiments of Turkish cavalry and a battery of artillery between them and their own lines. The order was passed for the Regiment to concentrate, as their extended line would be extremely vulnerable to a large enemy force. (However it was due to this extended line that the enemy gained the impression that the force was much larger than it was.) Because there was not yet sufficient sun to heliograph the situation to Brigade HQ, the 2nd LH fought a "skirmishing" battle with the enemy. It was during these clashes that Sergeant Carlyon and Lance-Corporal Blacket captured eleven Turkish Lancers, Sergeant Carlyon was awarded the Military Medal for his initiative and dash.

As soon as Divisional HQ was advised of the presence of the large Turkish force it ordered the 2nd LH Brigade to assist. In spite of the CO, Lieutenant Colonel Bourne, informing Brigade HQ, "Unlikely to require assistance . . . situation well in hand"[6]; Corps HQ became involved and turned out the entire Anzac and Australian Divisions who arrived after the battle was over. Later it was learnt that, under the cover of the Turkish force, German and Turkish Generals had been conducting a reconnaissance of the Australian positions in preparation for an attack.

The Charge at Beersheba

The arrival of the new Commander-in-Chief, General Allenby, inspired a cheery confidence among the troops of the 2nd LH. One of his first orders was to move his HQ close to the fighting, then he visited the troops. Gullet recalled that he "went through the hot, dusty camps in his army like a strong, fresh, reviving wind. He would dash up in his car to a Light Horse regiment, shake hands with a few officers, inspect hurriedly, but with a sure eye to good and bad points, the horses of perhaps a single squadron, and be gone in a few minutes, leaving a great trail of dust behind him."[7] The Light Horse loved him, the men affectionately called him "The Bull".

Allenby supported Chetwode's plan for a strike at Beersheba. He brought in reinforcements, more and better artillery, more aircraft and continued pushing the railway and pipeline eastwards. The key to the plan was to deceive the Turks into believing that the attack would be at Gaza. Artillery sites were prepared near Gaza, an army corps moved into position and a dummy railway terminal constructed. Offshore, warships prowled, and information was leaked to the Arabs that an amphibious landing would be taking place. In an elaborate ruse, phoney documents were "dropped" so that they would be found by the enemy. The authenticity of these was debated by the enemy, but the inclusion of letters from home, money, and blood stains on the haversack convinced the German intelligence officer that the documents were genuine. As a result the reserve enemy division stationed at Beersheba was hastily moved to Gaza.

The enemy anticipated that Beersheba would be attacked by a force of no more than one or two infantry brigades and cavalry from the south or west. Nevertheless the Turkish commander, Ismet, set about improving his defences on all fronts and made elaborate preparations for demolition. All the wells and water storages were wired with explosives, then demolition charges were placed in the ammunition dumps, railway carriages and locomotives. By late October, Beersheba represented a gigantic bomb – a deathtrap for any invader.

Beersheba, with its seventeen wells was well-defended by the Turks. Any army attempting to attack the town must do so across a vast area containing only limited water points. Water was to be crucial to the outcome of the battle. In the weeks preceding the attack the Light Horse, accompanied by Australian and New Zealand field engineers, made several sorties into the area to be used in the advance to Beersheba, surveying the water points at Khalasa and Asluj and enhancing their capacity as much as possible.

The attack was set for 31 October. A massive bombardment was concentrated on Gaza while the infantry moved out, in secret night marches, to prepare an attack on Beersheba from the south and west. Meanwhile the Desert Mounted Corps circled far to the south and moved in from the eastern desert flank. Beersheba's wells had to be captured intact and they had to be taken on the first day if the force was to survive.

The 2nd LH commenced the move to Beersheba on 24 October reaching Esani on the 27th. Enroute a squadron was detached from the Regiment to provide flank protection for the Desert Mounted Corps at Hill 840, six miles west of Beersheba, with orders to "dig in and hold it at all costs, in order to deny observation to the enemy".[8] This was only a temporary detachment and

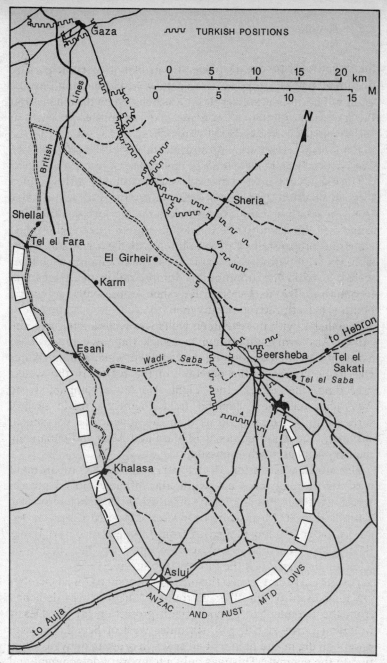

9 Beersheba, the wide sweeping move by the Anzac and Australian Mounted Divisions in preparation for the assault on the Turkish defences. (Time-Life Books Australia)

the squadron rejoined the Regiment prior to Beersheba. The Regiment paused at Asluj to water. Large parties had been at work improving the wells and building storage tanks but even so there was considerable difficulty in providing water for the men and horses of the entire Mounted Corps.

The 2nd LH, together with the remainder of the regiments of the 1st LH Brigade and the NZMR Brigade, had been assigned the task of securing Tel-el-Saba, a range of hills about three miles east of Beersheba. A frontal attack by the 3rd LH proved fruitless. About 2.00p.m. the 2nd LH was ordered to attack on the left of the 3rd Regiment. "We went in at the gallop and reached some mud huts about 800 yards from our objective. Here we dismounted and sent led horses back. Our coming in on the left of the 3rd Regiment drew the bulk of the defender's garrison that way, thus though we could not advance ourselves . . . it enabled the force on the right to push up to the Turkish position; but not before we sustained our heaviest casualty."[9] About 4.30p.m. an enemy shell killed the acting CO, Major Markwell. The loss of this gallant officer was irreplacable. The following is an extract from Routine Orders:

> The CO desires to place on record the severe loss to the Regiment, in the death of Major W.E. Markwell D.S.O. His loyal and devoted services to the Regiment since its formation were exemplary. His courage and energy in the field, his ability and conscientiousness as an administrator, his frank and generous nature, as a comrade, combined to mark him as one of the finest soldiers in the AIF, and his untimely death before reaching his prime, while in temporary command of the Regiment is a heavy blow to this unit in particular, and to the AIF as a whole.[10]

With the gaining of Tel-el-Saba, two troops of the Regiment were attached to the 1st LH — which was advancing dismounted towards Beersheba. The 1st LH had almost reached the town when the 4th Brigade was sent in at the gallop.

Despite the attack on Beersheba the Turkish commander, Ismet, could not convince von Kress that he was under a heavy assault. The German was still convinced that the major blow would fall at Gaza. Ismet was ordered to hold out but denied any reinforcements. Even so, with only 4,000 men and 28 field guns, he held off the attack by 58,500 men and 242 guns. Chauvel, stationed on a high ridge above the battle became increasingly concerned. As the afternoon shadows lengthened he appreciated that failure to secure the town by nightfall meant that the horses, many of which had not drink in nearly thirty-two hours (some forty-eight hours), faced a gruelling twelve hour march back to water at Khalasa and Asluj. Chauvel agreed with Brigadier Grant's suggestion that his brigade be permitted to attempt a cavalry charge on Beersheba.

At 4.30p.m., some twenty minutes before sunset, the 800 horsemen of the 4th and 12th LH Regiments of the 4th LH Brigade formed up and trotted towards the town. For the Turks it must have been a daunting sight, lines of cantering horsemen stretching twelve hundred yards across the broad, gently sloping valley. They were recognised as lighthorsemen and orders were given not to open fire until they halted and dismounted. At a distance of one and a half miles from the Turkish trenches the order to charge was given.

The horses began to gallop and the three waves of horsemen hurtled towards the Turks, bayonets flashing in the golden twilight. Turkish artillery immediately opened fire, joined by German aircraft and Turkish riflemen. The lighthorsemen swarmed over the Turkish trenches and towards the township.

While Chauvel and his staff had been considering the proposal to launch a cavalry charge, Ismet decided that further resistance would be futile and ordered a general withdrawal covered by a strong rearguard. At the same time he despatched the engineers "to destroy the water supply", thus denying a victory to Chauvel.[11] As the 4th LH Brigade galloped into Beersheba there was a terrific sustained roar of demolition explosions: ammunition dumps, a flour mill, a locomotive blown end over end, and the first two of the precious wells. However, the lightning attack and quick action by the Australians ensured that most of the wells and the reservoirs were captured intact.

With darkness falling, 58,500 men and 100,000 horses swarmed into Beersheba. It took almost 500,000 gallons of water to slake their thirst. The last great cavalry charge in history had saved the army with only 31 men killed, 36 wounded and the loss of 70 horses.

Gaza and Beyond

The successful attack at Beersheba had a dramatic effect on the enemy, which went far beyond Allenby's expectations. The fighting had been bitter and the Turks were not about to abandon the well-fortified town of Gaza without a struggle. Allenby's tactic was to keep the enemy guessing as to where the next blow was to fall: an attack on Gaza or a strike up the Hebron road to Jerusalem. When Chauvel detached a small group of 100 Arab auxiliaries (under the command of a British officer) to the north of Beersheba to harass the Turks as they withdrew, the German commander jumped to the conclusion that the British intended to advance to Jerusalem. Von Kress reacted to this move by withdrawing six battalions from the Gaza defences and hurled them against the minuscule unit, weakening the defence of Gaza. Chauvel's move had backfired as he needed the wells at Khuweilfeh, northeast of Beersheba, but the massive Turkish force blocked him.

The availability of water now became a limiting factor in Allenby's plans. The huge force, of almost 60,000 men and their horses, now in the Beersheba area was draining the supposedly unlimited water supply. To supplement the local supply a fleet of one hundred trucks, each with a 400 gallon tank, ferried water (day and night) from the head of the pipeline at Karm to Beersheba. Then a hot wind started to blow, drying up the few pools in the wadis, increasing the thirst of the men and horses. In choking dust, thousands of horses jostled around the wells at Beersheba. On 1 November Chauvel received Allenby's reluctant approval to move the entire Australian Mounted Division back to Karm and adequate water supplies. By the month of October, just before the early rains the area had been reduced to bare plains and harsh stony hills without a blade of grass. The already jaded horses were rationed down to a few pounds of pure grain. The planned attack by the Australians towards Gaza had to be postponed while the men and horses were given a chance to recover.

Pursuing the Enemy

The 2nd LH, which had concentrated in Beersheba, was delayed by a bombing raid on the Regiment's transport, resulting in twelve men killed, seven wounded, twenty-two horses killed, and the loss of a large quantity of stores and equipment. On 2 November the Regiment, with the remainder of the Brigade, moved out at midnight to attack the enemy now firmly entrenched at Khuweilfeh. The Turks guarding the only water in the area, were holding up the advance of the mounted troops in their sweep behind the Gaza line towards the coast. The 1st LH was leading the Brigade with orders to attack at 7.30a.m. A Squadron 2nd LH had been attached to the 1st LH to assist in the attack, while C Squadron was ordered to occupy a commanding rocky hill at Ras-el-Nagb – the key to the situation. Early in the afternoon the Regiment was ordered to concentrate for a Brigade attack scheduled to commence at 5.00p.m. However, C Squadron was under heavy attack and could not be relieved until 4.00p.m. and it was now apparent that the enemy was in much greater strength than first thought. The order to attack was cancelled and the Brigade withdrew to Beersheba to water. The 5th Yeomanry Brigade replaced the 1st LH Brigade and held the enemy until sufficient infantry and guns could be brought forward to dislodge the Turks. (A force of more than five divisions was eventually used to defeat the enemy at Khuweilfeh.)

On 6 November, although Khuweilfeh had still not fallen, it was decided to push towards the coast north of Gaza to capture the Turkish Army that was beginning to flee from Gaza. That night the Regiment, as the advance guard for the Brigade, rode to Sheria, about twelve miles, over a very rough track. Sheria had fallen during the night to the 60th Infantry Division and the Regiment was pushed forward to occupy an outpost line and maintain contact with the enemy. By first light shots were being exchanged with the Turkish cavalry and the Regiment was under artillery attack.

A two pronged advance was ordered by Chauvel: the Anzac Mounted Division striking for Jemmameh, the 60th Division towards Huj, while the Australian Mounted Division was to fill the gap between the two prongs. It was hoped that Chauvel's horsemen could cut off the enemy's retreat along the coast. The Turks, however, were fully aware of the danger of being trapped and were fighting a determined and effective rearguard action. Chauvel's thrust was delayed until the main body of the Turkish Army had made good its escape. Allenby and Chauvel had hoped that a large portion of the Turks could be overridden and forced to surrender, but this was not to be.

On the afternoon of 8 November the 1st LH Brigade succeeded in capturing Jemmameh. The village had a large reservoir and pumping equipment, so the Regiments were at last able to spend a relatively comfortable night, free of thirst and hunger. Over two hundred prisoners, two howitzers, machine-guns and a quantity of material had been captured. A determined counter-attack by an estimated 3,000 to 5,000 Turks was repelled by the 500 lighthorsemen.

The push to the coast continued and on 11 November the Regiment reconnoitered Wadi Sakerier and its surrounds. The plan was to secure the mouth of the wadi and the stone bridge that spanned it. It was an important wadi, the water ten feet deep and thirty yards wide, extended from Tel el Murre almost to the coast. As well as having ample supplies of good water, the area was a perfect site to stage the mounted brigades with a beach south of the wadi on which to land supplies brought in by ship. The 2nd LH advanced to the bridge with the 1st LH on their right. Despite strong Turkish resistance, during which the enemy gunners inflicted heavy casualties, the Australians succeeded in capturing and holding the wadi and the high ground dominating the area.

The Turkish Eighth Army continued its fighting retreat up the coast pursued by the Anzac Mounted Division as the enemy's Seventh Army withdrew into the Judean Hills towards Jerusalem, opposed by the Australian Mounted Division. On the coastal plain the men encountered fertile country and a number of Jewish villages. These were neat and orderly with attractive gardens, orchards and good supplies of food and wine. The people of the villages greeted the men as deliverers; this welcome, together with the prospect of capturing Jerusalem and

Jaffa, lightened the hearts of the battle-weary Troopers.

The Regiment took part in a variety of skirmishes and on 16 February entered Bethlehem. In the meantime the British and Australian Mounted Divisions had been fighting a determined Turkish enemy in the Judean hills. Judea was not a kind place for horsework, it was a jumbled pile of razorbacked ridges and narrow, rocky valleys, broken by groups of cone-shaped hills. At one time the rocky shelves, which jutted from every hillside, had held the cultivated soil of narrow terraces, but these had been washed away over the centuries leaving a harsh, barren landscape which both man and beast found difficult to negotiate. In addition the rain, wind and altitude combined to reduce the temperatures to near zero, causing numerous casualties from frostbite. Despite these obstacles the British forces succeeded in capturing Jerusalem on 9 December 1917.

The Push to the Jordan Valley

The Jordan Valley operations, the next major phase of the war in the Middle East, was the consequence of a directive from the British War Cabinet that Turkey must be completely eliminated from the war. Although for their part the British Generals believed that they had already brought the Turkish forces to a state of complete frustration with the capture of Jerusalem, the War Cabinet wanted the forces to concentrate on an offensive in Palestine which would completely remove Turkey from the war and open Germany's southern flank to attack.

The Egyptian Expeditionary Force, commanded by General Allenby, was by January 1918 camped near Jerusalem and Jaffa. Here, troops rested and reorganised while they awaited the next phase of the war. The advance of the broad-gauge railway, the repair of old Turkish lines and the landing of stores at Jaffa steadily increased the flow of supplies from the bases in Egypt. With the approach of spring and fine weather, General Allenby could plan on the maintenance of a substantial force in the Jordan Valley. By mid-February 1918, Allenby was ready to occupy the western side of the Valley from Wadi el Auja to the Dead Sea.

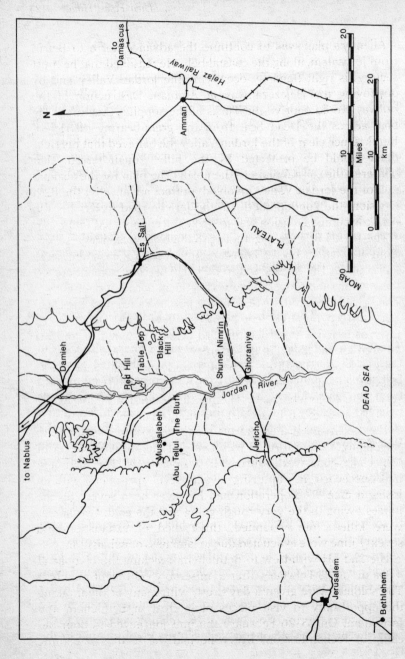

10 Jordan Valley, the scene of the Regiment's decisive battles at Ghoraniye and Abu Tellul.

Allenby's plan was to continue the advance north to Beirut from Jerusalem, along the coastal belt. He reasoned that he must secure his right flank by occupying the Jordan Valley and by destroying the Hejaz railway at Amman. Destruction of the railway would deprive the Turks of the supplies they drew, by boat across the Dead Sea, from the grain-bearing district of Kerak. Once clear of the Jordan Valley, he believed that his right flank would be protected by the hilly, rough terrain that bordered the coastal plain to the north. The plan for the occupation of the Jordan Valley involved a direct advance by the 60th Division (the Londoners) down the Jericho road to Jericho, the Anzac Mounted Division was to strike the enemy's left flank.

The 1st LH Brigade, together with the New Zealanders, set out from Bethlehem on 19 February into very rough country; in many places the horses had to be led. The New Zealanders led the advance and so were involved in the majority of the fighting. The Brigade reached the floor of the Jordan Valley near the Dead Sea, thirteen hundred feet below sea level, and struck north towards Jericho. A patrol of the 3rd LH and C Squadron 2nd LH entered Jericho ahead of the infantry. The Anzac Mounted Division had lost only three men in the advance, the bulk of the fighting having been undertaken by the infantry along the Jerusalem to Jericho road. The capture of Jericho allowed the establishment of a British line along the Jordan River.

The Regiment pushed patrols along the River, clashing with the enemy. Corporal Apelt was awarded the Military Medal for "especially good work" during these operations. The Regiment had succeeded in capturing twenty-seven prisoners without losing a man. The operation had however been severe on the horses owing to the very rough going and the cold: two horses were killed, four wounded, three died of exhaustion and seventy-nine were evacuated due to lameness or exhaustion.

The 2nd LH withdrew to Bethlehem, reaching the bivouac at 4.30a.m. on 23 February after a "miserably cold night march".[12] The soldiers were given a day's rest, with many of them taking the opportunity to visit places of interest in Bethlehem and Jerusalem. On 25–26 February the unit journeyed to their old camp at Richon-le-Zion. Everyone had the feeling of coming home, because the orange groves, flowering almond trees, hedged lanes, fields of bright flowers, and plantations of blue gums made a picture that reminded the men of home.

By 5 March, the Regiment had been reinforced with both men and horses and, after voting in the Queensland elections, was on the move to Beitin, north of Jerusalem. For the next two weeks the Brigade was held in reserve, conducting a number of patrols to the Jordan River. The weather was bitterly cold with continuous rain, so much so that the men were billeted – the first time this had been possible during the war.

Raid on Amman

General Allenby planned to launch an offensive as early as possible across the Jordan River to Amman. At Amman the railway passed over a long viaduct and through a tunnel which, if destroyed, would disrupt Turkish transport for several weeks, isolating the enemy to the south of Amman. The attack was under the command of General Shea of the 60th Infantry Division, who in addition to his own men was allocated the Anzac Mounted Division and Camel Brigade.

The rain interfered with the plans of the higher command. The Jordan, now in a state of flood, proved difficult for the engineers to bridge. However, after a night of feverish activity and against stiff Turkish opposition, the engineers of the Light Horse managed to throw a pontoon bridge across the torrent. On 23 March a bridgehead was established by the 60th Infantry Division who pushed the enemy well across the Valley. The following day the 2nd LH crossed the river and drove northwards forcing the Turks into the foothills. Despite enemy artillery and opposition from infantry outposts the men pushed up the precipitous goat tracks towards Es Salt and then northwards, to secure the high ground north of the Umm Es Shert.

The area was found to be occupied by the Turks and the CO ordered A Squadron to attack. To determine the enemy's strength a patrol was despatched to the main ridge where it discovered the enemy was in considerable strength, and being reinforced. The attack was abandoned. The Regiment withdrew to a stronger position, about a quarter of a mile to the south, where it could still prevent the enemy reaching the river. The Turks did not attack, moving the main force to the east, but left a strong group to guard the rear and prevent the 2LH from attacking. The Turks made no attempt to dislodge the Regiment — beyond constant sniping and artillery fire — which was providing valuable flank protection to the main force thrusting along the road to Amman.

At Amman the Anzac Mounted Division and 60th Infantry Division encountered 4,000 Turks and Asia Korps Germans dug in on high ground and supported by artillery and machine-guns. For three days the struggle waged. The Turks rushed in reinforcements by the train load. Soon the enemy were threatening General Shea's communications and beginning to isolate his force at Amman. Finally on 30 March, General Chetwode ordered General Shea to withdraw his force to the Jordan River. The wounded, carried on camel-borne cacolets or tied face down on horses, were evacuated first. Some of the Camel Corps troops froze to death and the horror of the march intensified as large numbers of refugees fled with the retreating columns. The force limped back over the Jordan, having achieved nothing at a cost of over one thousand casualties.

The Battle for Ghoraniye

A bridgehead was maintained on the eastern side of the Jordan at Ghoraniye by troops of the 180th Infantry Brigade. On 3 April the 1st LH Brigade, with the addition of the 5th LH, assumed responsibility for the defence of the bridgehead. The position had been organised into four equal battalion-size sectors. It was a surprise to the CO and men of the 2nd LH to find that they, the weakest Regiment numerically, had been given the two central battalion positions to occupy. The sector on the right was allocated to the 1st LH, the left to the 5th LH, while the 3rd LH was held in reserve.

Because of the length of the frontage, the CO placed all three

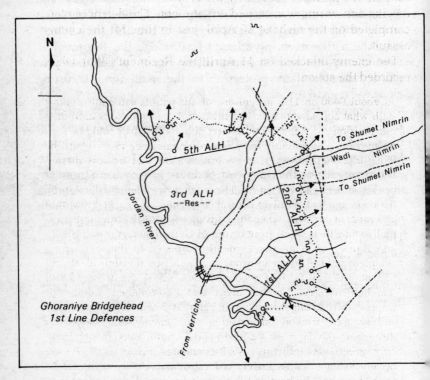

Frontage of Turkish attack on 11th April, 1918, along Wadi Nimrin from the East is shown thus

-Sketch by R.F. Bourne, Lieut R.E.

11 Ghoraniye Bridgehead, on the east bank of the Jordan River, was successfuly defended during the summer of 1918. It provided the launching base for the raids on Es Salt and Amman.

squadrons into the front-line. To make up for their numerical weakness the men worked like beavers to improve and consolidate the position. Due to enemy artillery the majority of the digging and wiring was carried out at night. The defences were completed on the night of 10 April, just in time for the enemy assault.

The enemy attacked on 11 April, the Regiment's War Diary recorded the action:

> At about 0430 on 11th instant one of our patrols came into contact with what appeared to be an enemy patrol . . . 1400 yards in front of our position . . . A few minutes later about 100 Turks could be seen advancing . . . apparently making a reconnaissance in force. At the light improved, however, it became evident that he was in considerably greater strength – many waves in extended formation could be seen advancing . . . The enemy was favoured by semi darkness and had excellent natural cover; and . . . reached to within 100 yards of our wire, when their advance was dealt with most effectually along their whole front by our M.G. [machine-gun] and Hotchkiss Rifles . . . Enemy was evidently surprised to find our wiring complete and fire power so strong . . . Enemy by this time [6.00a.m.] had moved 9 M.G. into good positions, raked our trenches and searched roads etc., in back areas; and a considerable number of snipers who had dug themselves in before it was light, kept up heavy and accurate fire on our position. The enemy also shelled us throughout the day with 4.2 [inch] and 77 [mm] guns, though these only succeeded in inflicting casualties amongst the horses . . . It soon became evident that the enemy had missed any chance of success he might have had, by waiting till it was light, as we were well able to stop any further advance by our small arms fire; and our artillery was dealing severely with his supports . . . At dark the enemy became active and it seemed possible that he would renew the attack – lights being seen and many voices heard in the Wadi at intervals till 0300 next morning. We kept touch by patrols who reported half hourly. At 0400 we raided and captured the nearest enemy posts; and our patrols following up, discovered that the main body had withdrawn; they pursued and captured some prisoners . . .
>
> Great credit is due to Captains S.N. McLean and W.J. Brown against whose posts the attack developed, for the manner in which they handled their squadrons . . . Lieutenant H.C. Kemp 1st L.H.M.G.S. [Light Horse Machine Gun Squadron] did splendid

work . . . The artillery co-operation was good and reflects great credit on the F.O.O.'s [forward observers] one of whom was mortally wounded. The following casualties were suffered by my regiment. Killed 6 O.R. [other ranks] Wounded 17 O.R., two of whom have since died. Enemy casualties on my front were as follows. Killed 151, actual count. Wounded and taken prisoners 25 O.R. Estimated wounded removed by enemy 500. Unwounded prisoners taken by this Regiment three officers and 63 O.R., 11 Arabs, eight Armenians . . .

We expended 38,000 rounds of . . . ammunition during the operation.[13]

It was Gullett's judgment that, "The highly successful defence was due in the main to the cool and complete control of the position from the outset by Bourne and his squadron-leaders McLean and Brown, to the habitual fine shooting of the light horsemen in a crisis, to the prompt and effective cooperation of the artillery, and to the admirable manner in which Lieutenant A.C. Kemp, of the machine-gun squadron, handled his guns".[14] The attack, which had been launched solely against the 2nd LH perimeter, boosted the men's confidence − confidence which was to be tested to the limit three months later. On 18 April the Ghoraniye Bridgehead was handed over to the 20th Indian Brigade, and the 1st LH Brigade moved into reserve (although it continued to provide patrols).

On 11 April at the same time as the attack at Ghoraniye, the Turks, in an effort to regain control of the Jordan Valley, also made a determined assault on Lieutenant Colonel Langley and his Camels at Musallabeh. From 4.00a.m. to 5.00a.m. the enemy shelled the position, his infantry creeping up the wadis in preparation for an assault. When the artillery barrage lifted, the Turks swarmed onto the position. As the Cameliers rose to fire many were shot by enemy riflemen stationed behind the assaulting waves.

For three hours the battle raged, some men even ran out of ammunition and hurled rocks down the hill at the Turks. Despite this, the Camels never lost control of the situation and by 8.00a.m. the attack petered out, although sniping and shelling continued throughout the day. The position was reinforced by a company of British infantry. In the late afternoon the Turks renewed their attack, but this was defeated by the combined fire of the Camels and the infantry; the enemy abandoned his efforts at nightfall. The next day Langley's men counted 170 enemy dead in front of their position. Victory had been achieved at the cost of 18 killed and 27 wounded. As at Ghoraniye, the Australians had won by the steadiness of leadership, good fire discipline and accurate shooting. The results of Ghoraniye and Musallabeh, coming after the failure of Amman, greatly raised the morale of all ranks.

The Es Salt Raid

While the assault on Amman had been underway, the Germans had launched their spring offensive in France pushing towards the channel ports. The War Office urgently ordered 60,000 men from Palestine to Europe. Reluctantly, Allenby surrendered two complete infantry divisions and twenty-two battalions. (Later he would be asked for a further division and half of the Light Horse.) He now had to set about rebuilding his army, using Indian infantry and cavalry, prior to his long-planned advance along the coast.

To give the advance every chance of success he once again launched an assault over the Jordan to deceive the Turks. On 18 April, in an effort to draw the Turkish reinforcements away from the coast and onto the plateau, he launched the feint against the Turkish stronghold at Shunet Nimrin in the foothills of the Moab. Although intended to be a feint, Allenby shortly afterwards ordered Chauvel to capture Shunet Nimrin and the town of Es Salt. Chauvel would be attacking a force of 8,500 Turks with the same force that had proved inadequate to defeat 4,000 Turks at Amman and Es Salt. The attack was to have been supported by 7,000 men promised by the King of Hejaz, but on the day of the assault the King consulted his oracle and was informed that the day was not propitious for a battle. The lack of cooperation by the Arabs had a disastrous effect on the outcome of the battle.

Chauvel's plan was to strike at Shunet Nimrin with the infantry and Anzac Mounted Division, while the Australian Mounted Division pushed up the east bank of the Jordan capturing Es Salt and so isolating the Turks at Shunet Nimrin. The operation met with early success, Es Salt being taken early on the second day by some reckless and dashing gallantry on the part of the Australians. The infantry, however, were having a difficult time at Shunet Nimrin. The enemy on the high ground of the plateau had wonderful artillery observation posts, enabling him to inflict severe casualties without being observed. As part of the Australian Mounted's drive to Es Salt the 2nd LH ascended the plateau, but the threat of enemy interdiction to the line of communication forced the Division to leave A Squadron to guard the road.

The enemy threat to the rear continued to grow and soon the entire Regiment was ordered to protect the rear of the Division to ensure that the mountain track was kept open. By the evening of 1 May, the enemy had succeeded in occupying defensive positions threatening the only withdrawal route. In a number of attacks against these positions the men excelled. In one assault Lieutenant King of Hughenden and twenty men were detailed to attack two posts at night. "The raid was brilliantly carried out — great credit being due to the leader — as owing to the very rough nature of the ground it was extremely difficult to locate the enemy in the dark, and surprise him. Five enemy were killed and four taken prisoner. The party also took seven horses, machine gun complete, 10 rifles, and 4,000 rounds SAA [small-arms ammunition]."[15] In addition a Turkish squadron supporting the posts fled in disorder. The attack had succeeded without a single casualty and for their outstanding performances Lieutenant King was awarded the Military Cross and Sergeant Geddes received the Military Medal.

Meanwhile, to the north, the Turks assembled a strong force and launched an assault on the 4th LH Brigade, located south of Damieh. The Brigade of 1,000 men were forced back by 6,000 Turks who assaulted in nine waves supported by artillery. The Brigade had a nightmarish withdrawal under pressure, being forced to abandon nine artillery pieces complete with breech blocks. To compound the problem the Turks launched a counterattack from Amman. After consultation with Allenby, Chauvel ordered a general retirement in order to save the mounted troops from being cut off by the Turkish push down the Jordan Valley. The 2nd LH was wholly occupied keeping open the only road to the Valley not held by the enemy. During the night 3-4 May, the entire force made a successful withdrawal from the plateau. The Regiment formed part of the rearguard, leapfrogging squadron by squadron towards the Jordan River. By early on the morning of 5 May the force had retired over the Jordan River.

A few days later the Regiment moved to Solomon's Pools, between Hebron and Bethlehem. The AIF Canteen and YMCA were established without delay and the men were able to live in comparative luxury. The stay at Solomon's Pools was devoted to training; there were lessons for potential NCOs, Hotchkiss gunners, signallers, and so on during the day and evenings. Each day a group of men, under the supervision of an officer, visited the Holy Places and areas of interest in and around Jerusalem and Bethlehem.

Birth of the 14th LH

The Es Salt operations marked the end of the career of the Imperial Camel Brigade. Under the capable leadership of Brigadier Smith V.C., M.C., the force had performed brilliantly at Magdhaba, Rafa, Gaza, and Musallabeh. After the penetration of the Gaza-Beersheba line, when the pace of the mounted troops became faster, the camels were easily outmarched by the horses. The camelier's greatest value had been in the western desert against the Senussi and in the desert wastes of the Sinai where they had fought as companies. Had the Brigade been created earlier it may have played a greater part in the campaign, but it was formed when the desert was already behind, and the men may have been of more value mounted on horses.

On the disbandment of the Corps the Australian Camel Companies moved to the Suez Canal (with their mounts). Here they were given horses and schooled afresh in handling the animals, the Australian Cameliers forming the 14th and 15th LH Regiments. A mock funeral was conducted and the "corpse", a flag covered camel saddle, was buried as Padre Houston conducted a burial service. The two newly formed Regiments shouldered rifles and fired a volley of shots. Many times during the Camel Corps existence the Cameliers had cursed their rough and often bad-tempered mounts, but now that these animals were lost to them, almost to a man, they felt real pangs of regret.

In 1920 a memorial to the men of the Imperial Camel Corps was unveiled on the Thames Embankment in London. It bears the names of the 346 members who lost their lives in Egypt, Sinai and Palestine during 1916-18; atop the plinth stands a bronze figure of a Camelier in complete marching order mounted on a camel.

The 14th and 15th LH Regiments formed the basis of the 5th LH Brigade under command of Brigadier G. Macarthur-Onslow. The New Zealanders of the Camel Corps supplied the machine-gun squadron of the new Brigade; the Australian element of the Camel Ambulance was mounted on horseback and transferred as a complete unit; and a few months later a French colonial regiment was added to the Brigade.

Defence of Abu Tellul

In June 1918 the 2nd LH together with the remainder of the Brigade returned to the Jordan Valley. The prospect of spending mid-summer thirteen hundred feet below sea level, where the temperature reached 128 degrees fahrenheit and malaria was rife, was not a pleasant one. The Brigade assumed responsibility for the Auju Sector from the 4th Brigade. This sector included the Musallabeh Salient, already made famous by the exploits of the Camel Corps, and which was to be the scene for the Regiment's most critical battle.

Initially the 1st and 3rd LH Regiments held the line with the 2nd LH in reserve. The Regiment kept its horses as it was appreciated that, in the event of an attack, support to the forward regiments would be required quickly. Each night one squadron was sent forward to assist with the wiring and entrenching of the defences. The bivouac area was shelled daily despite taking the greatest care to conceal the horses. On 30 June the 2nd LH relieved the 1st LH in the right sector of the Brigade position. The Regiment moved into the line with less than 230 men (supplemented by one section of machine-guns) having lost many soldiers due to malaria.

Abu Tellul was a dominating feature forming a salient which, had the enemy captured it, would have made the Ghoraniye Bridgehead over the Jordan untenable. A similar salient was held by the 2nd LH Brigade some three miles to the east of Abu Tellul. To the west some three to four miles away rose the mountains forming the western side of the Jordan Valley. These hills were held by the Turks and afforded them with excellent observation of the Brigade's position and movements. By employing their long-range guns from the hills and their field-guns immediately to the north, the enemy was able to shell the position almost continuously.

The features at Abu Tellul were so rocky that digging was difficult and in some places impossible. Sangars (rifle-pits) made of rocks were built above-ground in most cases instead of trenches. All movements, such as collecting rations, evacuating the wounded, erecting wire defences and sangars, had to be conducted at night due to the enemy's observation from the hills. The sangars were to prove useful against rifle and machine-gun fire, but not so effective against the shell splinters whizzing around in all directions.

The key to the position was Abu Tellul Ridge, divided for convenience into right and left sectors. At the first sign of an attack it was understood that the Reserve Regiment (now 1st LH) would immediately occupy the feature, a point emphasised by General Chauvel when he inspected the defences. Each night, assisted by soldiers from the Reserve Regiment, development of the position continued.

On 4 July the enemy raided Musallabeh but only succeeded in driving in the listening post. The enemy received a warm reception and the listening post was re-established, two prisoners being captured in the process. On the 6th another raid was driven off. The Regiment sent mounted patrols out to deny the enemy observation of the position and to raid the enemy's defences. From 7 July onwards the shelling intensified to 200-300 rounds per day, foreshadowing a possible attack. In preparation for an attack, additional water supplies in empty petrol-cans were stored in each post and the CO authorised an emergency supply of two bottles of beer per man — this was much appreciated.

On 13 July considerable enemy activity was observed to the north. Divisional HQ advised that it was believed that the enemy was preparing to withdraw. If that was the case the Turks obviously didn't intend to carry away any ammunition as the artillery fire increased! During the afternoon several new tents were erected at the enemy hospitals indicating that a major battle was imminent. The CO, Lieutenant Colonel Bourne, suggested to Brigade HQ that a squadron of the Reserve Regiment occupy its battle position on Abu Tellul. The proposal was rejected.

For the attack the enemy employed some of his best troops — three battalions of German infantry, the 702nd, the 703rd and the 11th Jaegers, together with the crack 24th Turkish Division. Two more German companies together with Turkish infantry attacked the 2nd LH Brigade to the east. The Regiment would at last be fighting the "real" enemy from that "real" and distant war in France. The men and officers had complete faith in each other. It was anticipated when the assault began, that the line would be pierced and the enemy would stream between the posts and Abu Tellul. It was confidently believed that the little garrisons would, although isolated, stand firm against the enemy for several hours until the Reserve Regiment arrived on Abu Tellul and crushed the enemy.

The Regiment occupied Mussallabeh with C Squadron plus one troop from B Squadron under Captain M.D. McDougall M.C. of Warwick, Maskerah with B Squadron (minus) under Captain F. Evans M.C., and Vyse with A Squadron (minus) under Major W.J. Brown. An advanced bombing post had been established at Vane, forward of the main defences, with orders

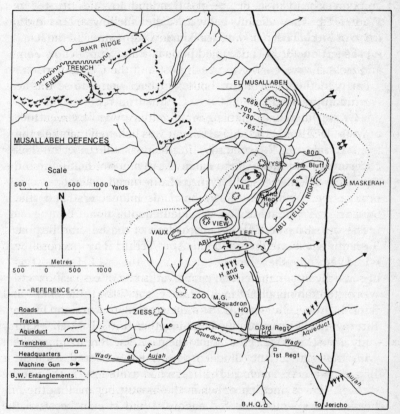

12 Musallabeh and Abu Tellul defences 14 July 1918. The 24th Turkish Division and three German infantry battalions tried unsuccessfully to dislodge the 1st Light Horse Brigade in some of the most desperate fighting of the campaign.

to fall back to Vyse if heavily attacked or at first light. Bourne kept three troops in reserve, one of which he placed on The Bluff and two close to Vyse. Regimental Headquarters was situated in Wadi Dhib, immediately behind Vale, which was held by a troop of 3rd LH currently under Bourne's command. To the Regiment's left, the 3rd LH occupied posts at Vale, View, Vaux, Zoo, and Ziess.

Early in the night enemy patrols were heard close to the Australian wire and, at 1.00a.m. C Squadron reported that the enemy appeared to be massing in the wadis immediately in front of Musallabeh. The 3rd LH also heard enemy movement to the front of Vale and View shortly after heavy shelling began; thus covering the approach of the enemy as they moved closer to the defences. The heavy artillery fire also severed telephone communications between Regimental Headquarters and the Squadrons, and rearwards to Brigade Headquarters.

The intensive artillery bombardment ceased abruptly at 3.30a.m. when one thousand enemy swarmed into Vale and the wired gaps on either side. The troop from the 3rd LH, consisting of some twenty men, had no chance of holding the position and was ordered by Bourne to withdraw to one of the prepared positions on Abu Tellul left. At the same time he sent a troop of the Regimental Reserve, under Lieutenant W.R. King M.C., to occupy a post on Abu Tellul Right; their orders were to cooperate with the post on the Bluff in delaying the enemy until the Brigade Reserve arrived. The enemy swarmed over Vale and through the ravines on either side – sweeping over Bourne's headquarters – then veered eastward between the posts and Abu Tellul. Bourne, with his staff, withdrew to Abu Tellul firing briskly as they went at the enemy who were hard on their heels.

At the same time, C Squadron at Musallabeh was being heavily attacked. Cutting through the wire entanglements the Germans bombed the occupants out of one of the posts, but an immediate counterattack led by Sergeant Carlyon M.M. drove them out. Although they attacked the four posts on Musallabeh many times, the enemy was driven back each time. The bombing party at Vane withdrew to A Squadron's position at Vyse at 2.30a.m. where the combined force withstood a heavy frontal attack, but was soon completely surrounded.

By daylight the posts at Musallabeh and Vyse were isolated and under vigorous attack. The men were not flustered and, having good cover, were able to bring effective fire to bear on the enemy now massed at their rear. As the Germans began to climb the northern slopes of Abu Tellul and The Bluff they were exposed to the machine-gun and rifle fire of Musallabeh and Vyse, as well as enfilade fire from the 3rd LH troops located at View and Vaux. The combined fire from these posts inflicted heavy casualties on the enemy.

B Squadron at Maskerah now came under heavy assault from the rear. Captain Evans, realising he would not be able to defeat the attack, rushed his men to a more suitable prepared, alternative position; here they could hold their ground and provide supporting fire to The Bluff. At 6.00a.m. the enemy attempted to advance down the flat to the north of Maskerah, but were dispersed by the fire from B Squadron.

Despite the heavy fire the Germans continued to climb the slopes of Abu Tellul and The Bluff and the situation became critical. The Bluff at this time was occupied by one troop, less than twenty men, under Lieutenant L.J. Henderson M.C. of Jimboomba, and the extreme right of Abu Tellul was held by Lieutenant King with a similar force. The two adjoining posts on Abu Tellul were manned by Bourne's regimental staff and a handful of men. As Bourne describes:

> These positions were obviously of vital importance, as even temporary possession of them by the enemy would have meant the putting out of action, if not capturing, of the three batteries situated immediately to the south of Abu Tellul. Further, had the ridge been taken by the enemy, it is doubtful whether the 1st Regiment would have been strong enough to retake it, and in the meantime the right flank of the 3rd Regiment would have been seriously threatened. My instructions to the troops holding them therefore, were to hold them at all costs till the 1st Regiment could come up.[16]

As day broke the Germans attacked all of these little posts employing over one thousand men under the cover of heavy and accurate artillery fire. Lieutenant King put up a most heroic fight but was overwhelmed by the superior numbers. He was killed directing his men, but his gallant troop fought on and the Germans did not enter the post until every man of the troop was killed or wounded. Henderson's men on The Bluff showed the same tenacity. Their young leader fell, seriously wounded, but he continued to direct the fight. His troop was reduced to only three men, but with this heroic element he held The Bluff against incredible odds until the counterattack relieved them.

Meanwhile the two posts further west on Abu Tellul were under fierce attack. Here the batmen, grooms, signallers from Regimental Headquarters, under Lieutenants G.T. Pledger (Adjutant) H.S. Wright (acting Signal Officer) and Sinton (Assistant Adjutant), were now the only troops between the Germans and the British batteries a few hundred yards behind. One of the posts was lost, and Bourne, who was personally directing the fight, withdrew the remnants to a prepared trench about 150 yards behind the crest. The enemy presented good targets as they were silhouetted on the crest, and for more than an hour this little force arrested the advance.

The Brigade reserve, the 1st LH, had been standing to, ready to advance once the order was given. Every man knew how critical this move must be to their friends in the 2nd and 3rd LH. At 3.40a.m. one squadron was ordered forward to report to Bourne on Abu Tellul. Shortly after 4.00a.m., when it was clear that the main blow was falling on the right sector of Tellul, Brigadier Cox sent a second squadron to join the first, at the same time the 3rd LH was ordered to send its reserve squadron to help the 2LH. Major Weir, who was leading the two reserve squadrons from the 1st LH, joined Bourne shortly after sunrise. His men had dismounted some time before and were now ready with fixed bayonets for the charge. With an eager shout the lighthorsemen topped the crest and charged at the Germans. Surprised and caught in the open, the enemy only offered slight resistance before fleeing down into the valley. As they ran they came close to the deadly fire of the defenders of Vyse and Musallabeh. Trapped and distracted they, "ran about like a lot of mad rabbits".[17]

This decisive stroke completely cleared Abu Tellul, but the enemy was still on The Bluff, where Henderson's men were fighting desperately. An attempt to reach The Bluff along the valley from the east was broken up by accurate enemy shell fire. With a subsequent attack by Major Weir, under artillery covering fire the lighthorsemen reached the beleaguered post about 8.00a.m. By 9.00a.m. all hostile forces within the perimeter had been accounted for and the original line restored.

The War Diary summed up the situation:

> The known casualties were: killed 55, wounded 45, prisoners 330. We collected 15 automatic rifles, 1 MG and 130 rifles. The performance was one of the best the Regiment ever put up – every man doing his duty efficiently under most strenuous circumstances. The heroic sacrifices of Lieutenant King and his troop deserve special notice. The gallant and stubborn resistance of Lieutenant Henderson and his troop rank amongst the best of the Regiment's performances. The steadiness of the three main posts and the able manner in which they were handled by their commanders (Major W.J. Brown, Captain M.D. McDougal and Captain F. Evans M.C.) were exemplary. The manner in which the Regimental details, under Lieutenants Pledger, Sinton and Wright, held their critical posts is also worthy of the highest praise – the complete success of the engagements being obviously due in great measure to the fact that the key to the position was denied till the Reserve Regiment arrived. Our casualties though serious, were very light considering the severity of the operation.[18]

The Brigade lost a total of 108 men but had soundly defeated the attack capturing 448 prisoners, of whom 377 were German, and inflicting heavy casualties upon the enemy. In five and a half hours of fighting the Brigade had expended 19,000 rounds rifle ammunition, 20,000 of Hotchkiss and 30,000 of machine-gun. The battle was the last deliberate offensive attempted by the enemy in Palestine and the only occasion in the campaign in which German infantry were used as storm troops. The news of the reverse had a damaging effect on the enemy's morale. To the Arabs from Jerusalem to Aleppo, the complete destruction of the German force did more to shake Turkish-German prestige than anything which had happened since the capture of Jerusalem.

The 15 July was employed in burying enemy dead and collecting the abandoned weapons and materials. Enemy parties, under the cover of the Red Cross, came out and collected their dead and wounded. The following day the war continued, with the enemy heavily shelling the positions and horse lines killing a further five men of the Regiment. That night the men were relieved by the 8th LH and the Regiment ordered out of the Jordan Valley for a rest. The Regiment was visited by Chauvel who congratulated the officers and men on their outstanding performance at Abu Tellul. Gullet, the Official Historian summed up the situation: "Bourne's forward leadership was cool and admirable, and the work of his 2nd Regiment of Queenslanders, on whom fell the brunt of the resistance, was never excelled in the career of the light horse."[19]

On 31 July the 1st Brigade was reviewed by the Commander-in-Chief General Sir Edmund Allenby. It was the Brigade's first mounted ceremonial parade since early 1915. Allenby addressed the men in highly complimentary terms, praising their work throughout the campaign, especially their performance at Musallabeh and Abu Tellul. For the next fortnight the Regiment attended classes, repaired and cleaned saddlery, organised sports meetings, and occasionally ventured to the beach for a swim. Then it was back to the Jordan Valley for the final stages of the war.

Final Days

The British now outnumbered the enemy two to one as Allenby prepared for the final phase. His opponent, the German General von Sanders who had commanded the defence of Gallipoli, had taken over a demoralised, neglected and poorly fed force. Eight of his ten divisions had been in the line without rest for six months, and they were losing over a thousand men a month from desertion alone. Nevertheless Allenby's offensive was planned in meticulous detail and brilliantly implemented by his commanders and men. He was to capture 370 miles of enemy territory in merely six weeks at a loss of only 5,000 casualties, when he and his staff had been planning for the loss of 30,000 men.

Chaytor, with his sick and tired Anzac Mounted Division, was to be left in the Jordan Valley together with a motley collection of infantry – British, West Indian, Jewish, and Indian units. The remainder of the force was to be assembled near Jaffa for an assault along the coast towards Damascus. To deceive the enemy into believing that a large force remained in the Jordan Valley new camps were established; 15,000 dummy horses built of canvas stood in lines (albeit quietly), camp fires were lit at night, sleighs drawn by mules jogged about creating dust, and the infantry marched into the valley in long open columns during the day only to be trucked out again under the cover of darkness. A dummy headquarters was established at Jerusalem and officers rented houses in the area for the summer. By the eve of the advance Allenby had reinforced his western front with two mounted divisions, one infantry division and 230 guns, yet the enemy believed that no major redeployment had occurred. A "Great Horse Show" and race meeting was advertised for 19 September at Jaffa. This ensured that the western sector could be regarded as a "dead" sector as the hundreds of horses taking part in the event were obviously not required for any major action. On the 19th, nearly the entire population of Jaffa turned out to watch the grand show. As they waited for "The Horse Show That Never Was", the Desert Mounted Corps was setting out on what was to become known as the "Great Ride".

For the final phase the bulk of the 2nd LH remained in the Jordan Valley, as part of the Anzac Mounted Division, for the offensive towards Amman. B Squadron, under the command of Major R.N. Franklin D.S.O., was honoured by being selected as the escort to the Corps Commander, Lieutenant General Sir Harry Chauvel of the Desert Mounted Corps, for the advance to Damascus. The 14th LH Regiment, as part of the 5th LH Brigade, was actively involved in the drive to Damascus. Let us follow the fortunes of each of these elements of the Regiments during the final days of the war.

2nd LH Towards Amman

The Anzac Mounted Division in the Jordan Valley was under

command of Major General Sir E.W.C. Chaytor. The Division had been strengthened by the addition of two battalions of West Indians, two battalions of Jews and an Indian brigade. The 1st LH Brigade had been designated as General Chaytor's reserve and the 2nd LH, due to its greatly reduced strength after the battles in June and July, malaria, and the detachment of B Squadron, was nominated Brigade reserve. The Anzacs crossed the Jordan on their third strike at Es Salt and Amman. Promises of support from Lawrence and the Arabs, which had failed to materialise on previous occasions, were to eventuate this time.

Although there were a number of short, vicious battles with the Turkish rearguard, the enemy, realising that there was a complete defeat of his forces in the west, did not offer as much resistance as previously encountered. On 23 September the 2nd LH, as part of the 1st LH Brigade attacked and captured Mafid Jozele. The Anzacs captured Es Salt and moved on Amman from the north and west.

The only sizeable Turkish force now east of Jordan was a 5,000 man garrison at Ziza. The Turks were dug in behind earthworks and surrounded by 10,000 Arabs. When the CO of the 5th LH, Lieutenant Colonel Cameron, rode into Ziza he was told by the Turkish commander that although they would surrender they feared that so few Australians would not be able to protect them from the vulturous Arabs. Cameron agreed and advised the Turks to continue manning their defences; the Arabs were told that if they attacked the Turks the Australians would attack them. Following the arrival of additional lighthorsemen, the 600 Australians, 5,000 Turks and 10,000 Arabs settled down for a strange night of comradeship. The Australians and Turks gathered around the fires, sharing their food, making chappaties together, demonstrating reciprocal respect and even admiration. When Arabs later attempted to creep past, the Turkish sentries replied with machine-gun and rifle fire – bringing bursts of laughter from the Australians and shouts of encouragement. The following morning after the arrival of more Anzacs, the Turks laid down their weapons and marched off into captivity.

This ended effective Turkish resistance east of the Jordan. The operations of Chaytor's force had resulted in the capture of

10,000 prisoners, 57 guns, 132 machine-guns, and a large quantity of valuable railway rolling stock. At the same time Chaytor's casualties had been amazingly light, 27 killed, 105 wounded and 7 missing. The 2nd LH was deployed to Suweile on 30 September and given the task of patrolling the area and ensuring the roads were kept open; there were still numerous enemy detachments in the mountains, although they were now without supplies or support. Malaria continued to ravage the Regiment so that by now there was only one man for every three horses. As far as the Anzac Mounted Division and the 2nd LH were concerned the fighting was over.

On 28 October there was a Ceremonial Parade for the raising of the Hejaz Flag at Amman, at which all units of the Anzac Mounted Division were represented. As a result of his cooperation the King of Hejaz was awarded control of the district of Amman. On the 30th the 2nd and 3rd LH Regiments were reviewed by Gaafar Pasha (a Hejaz officer) accompanied by General Chaytor. Gaafar Pasha had the unique distinction of wearing both the Iron Cross and the Companion of the Order of St Michael and St George (C.M.G.) — the former awarded when fighting with the Turks, the latter for his services to the British while under the command of the King of Hejaz.

The Great Ride

On the morning of 17 September Allenby outlined his plan for the coming attack. Security had been so tight that even the divisional commanders were not advised of the plan until two days before the advance was to begin. Allenby began his orders with a synopsis of the strategic situation: the Turks were entrenched across Palestine from the Mediterranean Sea to the Hejaz railway; they were expecting an attack in the east and were completely ignorant of the concentration of infantry and cavalry near the coast. The plan was to bombard the Turkish trenches just before dawn on 19 September, destroy the enemy's communications by air raids, force a gap with the infantry and then push the cavalry through, northwards up the coast. They would then move in behind the Turkish Seventh and Eighth Armies and trap them in the hills to be destroyed by the advancing infantry. Allenby concluded, "You have trained strenuously and devotedly at a time when you should have been enjoying a well-earned rest, after your long and trying summer in the Jordan Valley; but I hope and feel confident that you are at least about to reap the reward of your devotion."[20]

The 4th and 5th Divisions of Indian Cavalry and Yeomanry were to lead; the 4th striking towards von Sander's headquarters at Nazareth while the 5th cut through a pass and onto the Esdraelon Plain near Megiddo. The Light Horse would follow, hold the vital pass and aim for Jenin — a major enemy centre and key point on the escape route from Nablus.

The offensive began before dawn on 19 September when over three hundred guns opened fire on the sector of enemy trenches closest to the sea. The enemy was completely surprised. As the artillery barrage "rolled back" it was followed closely by the infantry who, within an hour, had penetrated the Turkish line.

The cavalry, 20,000 strong, crossed the old Turkish trenches shortly before dawn and then began the race northwards — a race that was to tax the fitness of both men and horses to the limit. It amazed the men to find that after crossing the first enemy line there were no redoubts or reserves in the rear, indicating that the deception in the Jordan Valley had been successful. By 11.00a.m. the force reached Liktera, on the coast, some twenty-five miles from the starting point. The vanguard rode hard and by the following morning had reached the Musmus Pass. Chauvel wrote to his wife, "I have had a glorious time. We have done a regular Jeb Stuart ride".[21] (General Stuart served in the Confederate Army in the American Civil War. A genius at using mounted troops as a scouting force, he was responsible for developing the Confederate cavalry into an effective mobile force. Regarded as a hero, Stuart was wounded on 11 May 1864 and died the following day at Richmond.)

The 3rd LH Brigade swung southeast to capture a large enemy force retreating towards Jenin. Groups of horsemen cut off the access roads to the town while a single squadron charged at a mass of Turks and Germans forming up in the town. The twenty-four yelling Australians captured over three thousand prisoners, many of whom had no idea that the front-line had been penetrated. By the next day the 10th LH had captured 8,107 prisoners including several divisional commanders and their staffs.

The ex-Cameliers of the 14th LH, under the command of Lieutenant Colonel G.F. Langley, were taking part in their first battle as lighthorsemen. The 14th LH were part of the 5th Australian Light Horse Brigade which was commanded by Brigadier-General G. Macarthur Onslow. The other regiments of the Brigade were the Australian 15th LH and a French colonial regiment. The 14th LH, together with the other regiments of the Australian Mounted Division, had recently been issued with swords in preparation for the offensive. Although training had been brief, the men practised for hours among themselves with the new weapon, astonishing their British instructors with the speed at which they mastered it. The secret of their success was due to their exceptional skill in the saddle and the tractability of their horses. When orders came for the Regiment to move from its position on the banks of the Auja where it, and other units of the 5th LH Brigade, had watered the horses, Langley recalled, "We, the old Cameliers, put on our shrapnel helmets, and started on our first operation as cavalry".[22] Shortly after this, the Brigade was accidentally bombed by British planes that had been bombing the retreating enemy. There were few casualties, but most of the transport and many animals were destroyed.

A little later a motor wagon and a car, with two Germans in each, took advantage of the confusion and tried to escape towards Nablus. However, Lieutenant Moseley, who was busy rounding up prisoners running for the hills, saw the vehicles move off and he galloped after the leading car. The vehicles soon became stuck in the debris on the road. The Germans jumped out and ran, pursued by Moseley who captured the first prisoner by the simple means of using a pistol butt to knock him unconscious. The other soldier promptly joined the fast-increasing column of prisoners. Meantime the motor wagon had also been stopped by the debris and Trooper Vigars used his sword to prevent the men's escape, wounding the driver in the shoulder. A group of would-be-escapees were surrounded by Lieutenant Cox and a few troopers, whose slashing swords soon convinced them to rejoin their fellow captives. Cox was killed shortly afterwards but before he fell he accounted for a number of the enemy. His sword, which was covered with blood from tip to hilt, was buried with him.

The 5th LH Brigade wheeled to the east with orders to capture Tul Keram and to cut the railway between Nablus and Jenin. Tul Keram was crowded with refugees and soldiers from the south, and the bombing by the airmen and the appearance of the Light Horse intensified the confusion. Thousands of troops swarmed east out of the village towards Nablus. The commander, Brigadier Macarthur-Onslow ordered his regiments to attack the flanks of the fleeing column; at the same time British and Australian airmen bombed and machine-gunned the column, creating havoc among the horse teams and vehicles in the narrow passes of Wadi esh Shair. By nightfall Macarthur-Onslow's Brigade had captured 2,000 prisoners, fifteen guns and large quantities of vehicles and stores. The Brigade was now scattered over several miles and it was not until 2.00a.m. on 20 September that it could be gathered together. At 7.00a.m. two squadrons of the 14th LH, under the sure leadership of Langley, succeeded in crossing the rough and trackless hills, they reached and destroyed a five mile section of the railway line.

The Brigade regrouped and advanced along the valley while the 3rd, 7th and 10th Infantry Divisions moved through the hills from the south. Small parties of Turks with machine-guns delayed the advance, but dismounted lighthorsemen swept wide on each flank to attack these positions from the rear. The Turks surrendered freely with their weapons intact! Although the countryside close to the Nablus provided good protection, the defence by the Turks was half-hearted. The town's civil leaders surrendered to the Brigade, handing the keys of the city to the Brigade-Major as the Light Horse rounded up 900 Turks. The Brigade was then ordered back to Jenin to rejoin the Australian Mounted Division. The Australians of the two newly formed regiments, together with the Frenchmen, had shown the same zest for the offensive that they had displayed in the Camel Brigade.

Valley of Death

The 5th Brigade moving along the road to Jenin swept before it the remnants of the Turkish force. The only way for the Turks to escape was by northeastern tracks across the mountains to Beisan and Jisr ed Damieh. Gullett described the slaughter, "Grim disaster awaited them. Descending to within a few hundred feet of their helpless quarry, the airmen quickly smashed up the leading vehicles and choked the gorge. Then flying up and down the doomed, chaotic train of motors, guns, and horse transport, through which surged thousands of distracted troops, the pilots and observers continued their terrible work with both bombs and machine-guns."[23] Probably nowhere else in the war was the devastating effectiveness of the air force against troops on the ground so convincingly demonstrated. Major O. Hogue, of the 14th LH, wrote, "Most of us had seen death and slaughter before. Some of us had four years of warfare behind us. But none of us will forget the day we rode through 'The Valley of Death'."[24] Within a week he would be involved in creating an even more devastating Valley of Death.

As the prisoners streamed in, the guard duties for the Escort Squadron (B Squadron 2nd LH) began to multiply. By the fourth day of the offensive they were guarding 12,000 prisoners with only two troops, half of the Squadron had already been sent rearwards as an escort for 4,000 prisoners captured earlier. Many of the prisoners had been marched from Jenin and were nearly mad with thirst. B Squadron marched them in batches to canvas troughs where they buried their heads in the water. Some filled their boots with water to take back to the compound.

Off to Damascus

On 22 September, Haifa was captured and within two days the magnitude of the victory began to be apparent, the Turkish Seventh and Eighth Armies had virtually ceased to exist. With the capture of Tiberias and Semakh on the Sea of Galilee, and Haifa and Acre on the coast, Allenby confided in Chauvel his desire to continue the advance to Damascus. The plan Chauvel devised was for the 4th Cavalry Division, in cooperation with Lawrence and his Arabs, to intercept the Turkish Fourth Army at Deraa while the Australian Mounted Division was to skirt to the north of the Sea of Galilee and march direct to Damascus. The 4th Cavalry Division was checked twice by Turkish rearguards before reaching Deraa, where the Arabs were already enjoying an orgy of looting and savage revenge among the enemy wounded. The divisional commander, General Barrow put an immediate end to the Arab savagery. The Division then swung north along the Pilgrims Road (a barely discernable track) chasing the Turkish Fourth Army as it fled to Damascus.

At the Jordan River the advance of the Australian Mounted Division was delayed by the discovery that the stone bridge had been destroyed, and the river was now defended by Germans manning machine-gun posts. The advance was halted for several hours until the 4th and 14th LH Regiments located a ford two miles south of the bridge and, under heavy enemy fire, scrambled across the stream and up the steep river bank. During the advance the Light Horse overpowered numerous enemy positions. The two Beersheba Regiments, the 4th and 12th LH, joined forces again to charge a superior enemy position at Kaukab causing the force of some two and a half thousand Turks to flee without firing a shot.

As the Australians approached Damascus, the 5th LH Brigade moved to the west of Daraya along the foothills of the Anti-Lebanons. Here the enemy made a brief stand, but Langley, with the 14th LH, cleverly outflanked them. The enemy was utterly demoralised and, as the Regiment opened fire across their rear, some five to six thousand Turks fled in confusion towards the Barada Gorge and onto the road that led to Beirut.

The gorge, as it winds between the sheer cliffs, is no wider than one hundred yards. Along this confined passage, crossing from side to side, is the roaring Barada River, and crowded along its sides run the road and the railway. The Frenchmen of the Brigade swung into the hills, gained the heights above the road and railway in the Gorge, and engaged the head of the fleeing enemy column. At the same time a handful of men, under the command of Major Hogue, occupied a house at the entrance to the gorge and poured fire into the Turkish rear. The enemy was trapped! They threw down their arms and surrendered, over four thousand prisoners were captured. This dashing episode was only the beginning of the tragedy that was to befall the Turks in the Barada Gorge.

Chauvel's plans called for the city to be isolated by seizing the Barada Gorge and the road to Homs, meanwhile the remaining divisions were to attack Damascus from the south. The 3rd and 5th LH Brigades moved through the steep, rough hills and by mid-afternoon occupied the heights along some four miles of the Barada Gorge. As the men looked down they saw a narrow floor crowded with great columns of fugitive troops, refugees, transport, and crowded railway trains fleeing Damascus.

The official history describes the action:

> German machine-gunners, operating from the tops of motor-lorries and trains, defied the challenge to surrender, and all along the gorge the unequal issue was joined. The result was sheer slaughter. The light horsemen, firing with fearful accuracy, shot the column to a standstill and then to silence. For miles the bed of the gorge was a shambles of Turks and Germans, camels and horses and mules. Never in the campaign had the machine-gunners found such a target.[25]

It would take 300 German prisoners more than two weeks to clear the debris.

With orders not to enter Damascus, the men had a grandstand view as Turks burnt stores and blew up ammunition dumps. Major Olden, of the 10th LH, while patrolling the outskirts of the town early on the morning of 1 October, noticed a huge crowd outside the Hall of Government and went to investigate. Swords in hand, the Australians clattered over the bridge, charged through the crowd, and pulled up in front of the building. Olden drew his pistol and demanded to see the civil governor. He was escorted inside and presented to Emir Said who told Olden that he had been installed as Governor before the Turkish commander fled the city the previous day. Emir Said surrendered the city of Damascus to "the first of the British Army".[26] He promised

Major Olden there would be no more shooting in the streets and formally wrote out his assurance to commemorate the historic occasion. On hearing the news the crowd became delirious with excitement; as the Australians left they were cheered, applauded, kissed, and showered with flowers and rose water — it was a taste of victory that few modern soldiers have shared.

The honour of officially "capturing" Damascus was reserved, for political reasons, for Lawrence of Arabia and his Arabs. Dressed in his white and gold robes, riding in a blue Rolls Royce, Lawrence entered the city in triumph some two hours later. The city was in turmoil. The Arab despised the Christian; the Christian feared the Arab and both hated the Turk. After 400 years of Turkish power a vacuum had been created with the departure of the Turks. Most of the local notables were corrupt or incompetent, and although many of the Arab civil servants remained at their posts, the strong guiding hand of the Turk had gone. Lawrence had been followed into Damascus by thousands of Arab horsemen, armed to excess with rifles, swords, daggers, and revolvers. Most had provided service in return for gold and the prospect of loot. Behind the Arabs came Bedouins, lured by the splendid prospect of looting the town. All day they loaded their women and animals with staggering amounts of booty.

Chauvel decided to calm the situation by a show of strength. Escorted by the B Squadron 2nd LH (under command of Major Franklin) Chauvel rode through the city at noon on 2 October followed by units of his three divisions. All the might of the British Empire was represented: Yeoman and gunners from the UK, lighthorsemen from Australia and New Zealand, lancers from India (as well as cavalry from France). The effect of the spectacle on the population was electric. The great prancing horses, grim tired men with swords and lances, and the guns rattling over the cobblestones struck fear and order into the dense masses of people lining the streets. The fanatical excitement vanished and within a few hours the city returned to normal.

Chauvel assumed responsibility for Damascus. He had the task of finding supplies and forage for his Corps with its 25,000 horses; there were now 300,000 Damascenes who must be fed and whose government must be put in order, and about 20,000

Musallabeh, the scene of heavy fighting, first involving the Imperial Camel Corps in April 1918 and later the 2nd Light Horse Regiment in July 1918. (Australian War Memorial)

Musallabeh or "Camel's Hump", the scene of the attack against the 1st Light Horse Brigade. Major C. Stodart and his batman Trooper J. Fowler of the 2nd Light Horse Regiment are sitting with their backs to the camera. (Australian War Memorial)

Mounted and dismounted Turkish soldiers await the enemy near Jaffa, September 1918. In the distance can be seen the dust from the approaching horsemen. (Australian War Memorial)

Disaster awaited the Turks as they fled along the narrow gorges. The airmen were able to shoot the enemy columns to a halt while the horsemen completed the devastation. Major Hogue of the 14th Light Horse Regiment described it as riding into a "valley of death". (2/14 Light Horse Archives)

The scene in the Barada Gorge, October 1918. In this gorge large numbers of German and Turkish troops who were retreating from Damascus were slaughtered by the lighthorsemen when they refused to surrender. (Australian War Memorial)

Presentation of the 2nd Light Horse Regiment Guidon at Enoggera on Sunday 18 March 1928. (Courtesy of L. Friend)

Colonel P.J. Bailey, Commander of the 1st Cavalry Brigade, presenting the Guidon to the 14th Light Horse Regiment at Enoggera on Sunday 18 March 1928. *(Daily Mail)*

The Harrisville Troop with the Prince of Wales Cup and the Lord Lamington Shield. Under the command of Lieutenant Bob Coppin, the Troop was Queensland's premier Troop for three consecutive years, 1931, '32 and '33. (Courtesy of C. Head)

The Boonah Troop on the march in the 1930s. (Courtesy of R. Warren)

Toogoolawah Troop's tent-pegging team at the Toogoolawah Show in 1938. (2/14 Light Horse Archives)

Laidley Troop during a three month camp at Beaudesert in 1940. Included here are: D. Clarke, A. Clarke, A. Lester, C. Hermann, R. Moller, R. Kimlin, C. Peacock, J. Manthey, B. Whitehouse, and L. Noyes. (Courtesy of Max Upham)

The Headquarters Medium Machine Gun Troop, March 1940. (Courtesy of Max Upham)

A troop involved in a Prince of Wales Cup competition at the Enoggera range. In the foregound is the Hotchkiss machine-gun team providing fire support to the riflemen in the distance. (Courtesy of L. Friend)

1st Cavalry Brigade being reviewed by the Governor-General Lord Gowrie in March 1940. (Courtesy of Max Upham)

The last ceremonial ride-past of the 2/14 Light Horse Regiment. The parade was held at Southport in 1940 and reviewed by the Governor-General Lord Gowrie. In the foreground is Syd Appleby leading the Boonah Troop. (Courtesy of G.H. Studios)

George Reid and his mount on the beach at Southport 1940. (Courtesy of the Reid family)

(Courtesy of T. Huchinson)

Watering lines at Black River, North Queensland, in 1942. (Courtesy of K. Gregory)

Tony Cousins and Ben Wall of "Yorkforce", at Alligator Creek, North Queensland in 1942. (Courtesy of K. Gregory)

The Regiment driving through Queen Street in Brisbane during the Coral Sea Parade in 1950. At this time the unit was equipped with

The Australian Prime Minister's escort to the coronation of Queen Elizabeth II in Hyde Park London, 1953. From left, Captain N. Park, Warrant Officer D. Piefke, Sergeant R. Birks, and Warrant Officer A. Martin. (Courtesy of N.R. Grinyer)

Reconnaissance Troop, Headquarters Squadron, winners of the Prince of Wales Cup in 1953. (Courtesy of L. Friend)

106mm recoilless rifle in action at Tin Can Bay during the 1973 annual camp. The 106mm RCL was an anti-tank weapon with an effective range of 1200 yards but was hampered by a large signature when fired. (2/14 Light Horse Archives)

An M113A1 being off-loaded at Bulimba following its return from an exercise. Th American-built vehicle weighs 11 tons and is equipped with .30 and .50 inch machine guns. It has a crew of two and is capable of carrying eleven infantrymen. (2/14 Ligh Horse Archives)

The Honorary Colonel Bill Morgan, Lieutenant Neville and Trooper Gardiner sharing a joke at the annual camp at Tin Can Bay in 1973. (2/14 Light Horse Archives)

Lieutenant Colonel Tom Childs presenting Lieutenant Steve Eastaughffe with the Polo Cup for the best troop during "Emu's Revenge", September 1985, at Shoalwater Bay. (2/14 Light Horse Archives)

Trooper F.C. Duce, a lighthorsemen of the Anzac Mounted Division, on Anzac D 1983. *(Sunday Sun)*

Turkish prisoners to be cared for. Dysentry, cholera, typhus, influenza, and malaria were rampant. Chauvel set his staff and men to work: clearing the streets of dead Turks and horses, attending the 1,800 patients lying unattended in hospital, organising food distribution, policing the city and re-establishing the public service infrastructure. Following some confusion in the initial few days, Damascus gradually assumed an air of normality.

However, Chauvel was soon confronted by a new problem when a sudden surge in the sickness rate began to decimate his Corps. The malignant malaria, picked up during the offensive in the Jordan Valley, now struck down hundreds of men — over twelve hundred were admitted to hospital in the week ending 5 October. Then on 6 October, the world-wide epidemic of pneumonic influenza descended upon the Desert Mounted Corps, with the result that within one week over 3,100 were hospitalised. Chauvel's command was dwindling under his very eyes. Among the casualties were Sergeant Linan and Trooper Sinclair, both of B Squadron 2nd LH, who died of illness during the halt at Damascus.

Armistice

The Desert Corps had achieved much during the twelve day advance to Damascus: the Light Horse Regiments had travelled 250 miles as the crow flies, but many had ridden nearly 400 miles, the Australian Mounted Division had captured 31,335 prisoners for the loss of only 92 men. The Turkish Army had all but ceased to exist.

On 5 October 1918 the 4th and 5th Cavalry Divisions set off north in pursuit of the enemy while the Australians remained to guard Damascus. Chauvel was forced to call a halt to the advance after a few days because of the appalling losses due to illness, some brigades losing forty to sixty per cent of their men in one week. The tired and sick Australian Mounted Division was called out of Damascus to replace the 4th and 5th Cavalry Divisions. At Homs, northeast of Tripoli, the lighthorsemen learnt of the armistice that had been declared at noon on 31 October — one year to the day after the charge at Beersheba. The war was finally over.

The news was an anti-climax. The absence of excitement was partially due to the fact that the armistice had been expected, but it was mainly due to the mental and physical weariness of the men. The epidemics at Damascus had not only caused many deaths, but had shaken everyone's nerve and left the regiments depressed and weary beyond expression.

The voluntary service of the Australians was deemed to be at an end and the policing of the captured areas was allotted to the Indian and British troops, whose tour of duty in Palestine had been relatively brief. The 14th LH, together with the remainder of the Australian Mounted Division, marched from Homs to Tripoli where they went into a well-supplied and comfortable camp while awaiting transport to Australia.

Meanwhile the Anzac Mounted Division, after its swift and decisive offensive from the Jordan to Amman, was withdrawn first to Jerusalem and then to Richon. Chaytor's men also suffered severely from disease. The fighting at Amman was scarcely over when malaria and influenza swept through the ranks. The Australians and New Zealanders were especially afflicted; it was the 1st, 2nd and 3rd LH Brigades together with the New Zealanders who, alone of all of the troops in Allenby's command, had been subjected to all of the rigours of the campaign since the first crossing of the Canal. Other Light Horse regiments had been involved for as long, but not in actual contact with the enemy.

From April 1916 to the end of the war, none had a record like the famous Anzac Mounted Division, first commanded by Chauvel and later Chaytor. Fighting at Romani, Magdhaba and Rafah almost single-handed, they had been responsible for clearing the Sinai Peninsula; they had fought at Beersheba; they had been the first troops into the Jordan Valley and the last to leave that sinister area. For the final advance they had been in an even poorer physical condition than the men in the Australian Mounted Division. So severe was disease that at the close of the campaign there were 900 stretcher cases at Jericho alone. The regiments were so decimated that in the withdrawal from Amman the riderless horses, for the first time in the campaign, were driven along the tracks in mobs.

The thoroughness of the victory was a tribute to Allenby, his commanders (notably Chauvel) and to excellent staff work. Within one month three enemy armies had been destroyed, 75,000 prisoners captured, 360 guns, 800 machine-guns, 210 trucks, 50 cars, 90 railway engines, 470 freight cars, and 3,500 transport animals captured. By the middle of October there was scarcely a Turk to be seen between Jaffa and Aleppo.

The armistice was followed by the occupation of the Turkish capital. Sentiment prompted a decision to send an Anzac force to Gallipoli, while all regiments vied for the honour, the 7th LH and Canterbury Mounted rifles were chosen. The Anzacs spent six weeks at the task of locating the remains and graves of fellow soldiers, and collecting trophies which were later to be displayed in the Australian War Memorial.

Destroy the Horses

On 30 January demobilisation began. Original members of the Regiment who had urgent family or business reasons for wanting to go home were allowed to leave for Australia immediately. Only twenty-four men elected to do this. On 23 February 1919, the 2nd LH Regiment received orders to destroy all horses over the age of eight years. The younger animals were to be sold on the local market. Because of quarantine restrictions the horses could not be brought back to Australia, there was too great a danger that exotic diseases would be introduced. Nor was it possible to retain them at some site in Egypt.

The news that they would lose their faithful mounts was a blow to the lighthorsemen who had shared so much with their horses. These animals had given splendid service often under appalling conditions; with an average weight of twenty stone on their tired backs and often with little or no water to drink, many had travelled over ten thousand miles. Most troopers felt that a quick death by a merciful bullet was an end more befitting such comrades than the slow misery of life with the Egyptians, or some other strange riders who might ill-treat them.

Veterinary Officer G.E. Fethers later wrote of his sadness in having to classify horses for either destruction or hand over. Many troopers, preferring a quick death for their horses, tried to convince him that they were more than eight years old, even when date-brands and the animals' teeth told otherwise. Horses were classified A, B, C, and D according to age and condition. As a massive protest swept through the ranks the idea of selling walers was dropped, and orders given that A and B horses would be issued to the British cavalry units. C and D horses were to be shot, their manes and tails shorn for sale as horsehair, shoes removed for recycling, and the animals skinned.

A last race meeting was held before this took place. The horses were then led away to olive groves outside Tripoli, they were tethered in familiar horse lines, given a last nosebag of feed, and then shot by special squads of marksmen.

Just how many of these horses were quietly taken out by their owners and shot will probably never be known. Trooper Bluegum put into verse the feelings of the lighthorsemen in "The Horses Stay Behind":

> In days to come we'll wander west and cross the range again;
> We'll hear the bush birds singing in the green trees after rain;
> We'll canter through the Mitchell grass and breast the bracing wind;
> But we'll have other horses. Our chargers stay behind.
>
> Around the fire at night we'll yarn about old Sinai;
> We'll fight our battles o'er again; and as the days go by
> There'll be old mates to greet us. The bush girls will be kind;
> Still our thoughts will often wander to the horses left behind.
>
> I don't think I could stand the thought of my old fancy hack
> Just crawling round old Cairo with a 'Gyppo on his back.
> Perhaps some English tourist out in Palestine may find
> My broken-hearted waler with a wooden plough behind.
>
> I think I'd better shoot him and tell a little lie:
> 'He floundered in a wombat hole and then lay down to die.'
> Maybe I'll get court-martialled; but I'm damned if I'm inclined
> To go back to Australia and leave my horse behind.[27]

After a number of hitches, the handing over of animals to the Imperial Remount Service proceeded. This was almost complete when a disturbance broke out between the Australian and New Zealand troops and the Arabs at and around the village of Surafend.

Surafend

These natives, probably secure in the knowledge that the nearby Anzac camps would soon leave the area, grew audacious in their pilfering. Night after night the troopers lost property from their tents. The Australians and New Zealanders slept soundly, and were easy prey for the cunning, barefoot robbers of Surafend, who by this time had been "reinforced" by some nomad Bedouins camped near the village.

Trouble boiled over when a New Zealander was awakened by an Arab pulling at a bag which served as a pillow. The New Zealander shouted for help, leapt to his feet and pursued the thief through the camp. As he overtook the Arab, the man turned, shot and killed his pursuer. The camp was aroused by the shot and the New Zealanders chased the Arab over the loose sand to Surafend. They placed a cordon around the village to prevent anyone from leaving and waited for morning. At dawn the head men were summoned and ordered to surrender the murderer. The sheikhs were evasive and pleaded ignorance. During the day the matter was discussed with the staff at Divisional Headquarters, but by nightfall nothing had been resolved.

That night the New Zealanders joined by large groups of Australians, entered the village and, after evacuating the women and children, used heavy sticks to beat the men. They then burnt the village. Many Arabs were killed, and few escaped without injury. The village was demolished. Next the Anzacs raided and burnt the nearby Bedouin camp, before going quietly back to their tents.

Allenby, who had failed to take any disciplinary action against earlier petty thieving by the Arabs, was not nearly as lenient with the lighthorsemen. He arranged for the Anzac Mounted

Division to be paraded before him. Then he roused the ire of the Light Horse by a savage dressing-down. The men would have accepted disciplinary action over the raid at Surafend much more readily than the ill-considered language of their Commander-in-Chief. The final words in his denunciation were "I was proud of you once. I am proud of you no longer!"[28]

After that, the whole Division was moved to a desert camp near Rafah. All members of the Division who were on leave were recalled, and Allenby had the names of all officers and men of the Division withdrawn from the list of those recommended for honours and decorations. Feelings ran high, and it soon became apparent that unless someone forced the Commander-in-Chief to realise the harm being done to Imperial and Commonwealth relations, the Australians and New Zealanders would take home their resentment, which could cause irreparable damage to the Empire.

At this time Lieutenant H.S. Gullett, war correspondent, had just been appointed official historian of the part played by Australians in the Sinai and Palestine campaigns. He sought an interview with Allenby, and spoke forcefully to the Commander-in-Chief at Cairo Headquarters. Allenby had been nicknamed "The Bull" and he roared like one upon hearing of Gullett's reason for calling to see him. Gullett however was not intimidated. He pointed out to the enraged Allenby that he was now a civilian, and a representative of the Australian government to boot. He counselled Allenby to think very hard for a moment about what damage could be done to relations between Australia and New Zealand and the Mother Country if he refused to admit there was a problem to solve, instead of a punishment to be imposed.

After the Commander-in-Chief had simmered down he accepted Gullett's advice. He then wrote an Order of the Day, which was issued at once to every soldier concerned. As well, Allenby issued a personal tribute to the Light Horse which (in part) read:

> The Australian Light Horsemen combined with a splendid physique, a restless activity of mind. This mental quality renders him somewhat impatient of rigid and formal discipline, but it confers on

him the gift of adaptability, and this is the secret of much of his success, mounted or on foot. In this dual role, on every variety of ground; mountain, plain, desert, swamp or jungle, the Australian Light Horsemen has proved himself equal to the best. He has earned the gratitude of the Empire, and the admiration of the world.[29]

Home

On 10 March Major General Chaytor made a final inspection of the 2nd Regiment, thanked the men for their services throughout the campaign, and said goodbye. On 12 March, the Regiment boarded a train at Kantara, and with some members of the 1st Brigade Detail the 2nd LH Regiment embarked next day for Australia on H.M.T. *Ulimaroa*. The voyage home was uneventful, physical training and sports meetings occupied most of the time. There was a short stop at Colombo where Major Franklin was disembarked to be admitted to hospital. He reached Sydney two months later, but died there in a Military Hospital. There was something singularly tragic about the deaths of those who, having survived the horrors of war, were robbed of the joy of reunion with their families and friends.

The ship called at Fremantle, arriving at Sydney on 18 April where it was quarantined for four days. Forty men, who came from the Northern Rivers of New South Wales, were disembarked and the ship sailed on to Moreton Bay. Much to the disgust of the men they were again quarantined, this time for seven days at Lytton. Finally on 30 April 1919, they had the honour of marching through the streets of Brisbane, the State Governor taking the salute at Albert Square. Under the Regiment's first Commanding Officer, Colonel Stodart, a guard of honour, made up of returned men of the Regiment, formed at the saluting dias. The Regiment was dismissed at the drill hall in Adelaide Street and that afternoon the Regimental colours laid up in Saint John's Cathedral.

The Egyptian Rebellion

The Arabs had been richly rewarded for their part in the destruction of the Turks. Though their casualties had been few, and the hardships they had endured insignificant they had been well-paid in gold, weapons and munitions, and had been awarded large tracts of territory – including the rich and bountiful city of Damascus. The war had united the Arabs as never before and they now came together in a common cause.

Having overthrown the Turks, the spirit of freedom burned from the Mediterranean to the Euphrates, this augured ill for the Jews in Palestine, the French in Syria and the British in Mesopotamia; further, it added impetus to the smouldering unrest in Egypt. In mid-March 1919 the long-simmering rebellion in Egypt erupted. The uprising took the remaining Light Horse regiments, including the 14th LH, back to war.

At the time the Anzac Mounted Division (less the 1st and 2nd LH) was still at Rafa, whereas the Australian Mounted Division had been moved by sea from Tripoli to Moascar. All of the units had handed in their equipment and were awaiting embarkation for home. Because there were no large formations of British troops in Egypt to quell the rebellion all embarkation was cancelled, and within twenty-four hours the 3rd LH Brigade was on the march across the desert to Zagazig. The response of the troops was impressive; they abandoned without a murmur their hopes of an early return to Australia, and set out on their new enterprise with determination.

Soon all of the Anzac regiments (with the exception of the 1st and 2nd LH which had departed for home) were in the saddle, the area of operations extended from Upper Egypt to the Delta. So urgent was the need for mounted troops, that even the men convalescing after hospitalisation were drafted. There was no organised fighting, but several decisive brushes with the rioters cost the Australians twenty casualties while several hundred Egyptians died.

The Egyptians lost their nerve when they saw the lighthorsemen and were soon, under "guidance" from the soldiers, very actively engaged in repairing broken railways, and generally putting in order the places they had damaged. Within a month the danger was past. This time prior to embarkation, the mounted troops were comfortably billeted beside the Nile, where the abundance of fresh food and water ensured a very pleasant few weeks.

The efforts of the lighthorsemen in "keeping the peace operations" carried out in Egypt were recognised on 1 May 1919, when Lieutenant General Sir E.S. Bulfin KCB, CVO, Commanding the Force in Egypt, informed Brigadier General L.C. Wilson that he wished him to convey, to the troops under his command, his appreciation of the most valuable services rendered by the Australians during the recent disturbances. Further, that he recognised that they were his great standby in this crisis; that he had noted the very soldierly manner in which these services had been rendered; and that he was proud to have command of them.

On 1 July 1919, Major General P.C. Palin, Military Governor of the East Delta Area, wrote to Major A.S. Nobbs of the 14th LH:

> I want to thank you and your Regiment for the excellent work which you have done under my command since the outbreak of the disturbances in Egypt. The tactful manner in which situations have been handled by the Officers and the cheerful and willing conduct displayed by all other ranks, enabled the East Delta Area to be quickly brought to its normal state. The good discipline exhibited by the men under trying and often tempting circumstances, has filled me with admiration.
>
> It has in a very large measure, brought home to the Egyptians the enormity of their wrongdoing under the influence of those who led them astray. The impression left by you is a good one and I am sure will be a lasting one.
>
> On the eve of your departure to Australia, I wish you a pleasant and safe voyage and the best of luck in the future.[30]

On 11 July, the 14th LH boarded a train at Moascar on the first stage of a journey to Suez to board HMAT *Duncola*. At 5.00p.m. on 24 July 1919, the *Duncola* sailed for Australia, the men of the 14th LH Regiment – the gallant ex-Cameliers – were on their way home.

Historian Bill Gammage reached the following conclusion about the lighthorsemen:

> The fierce individuality with which he fought Turks, Arabs, and English staff officers lay close to the heart of the Australian light horsemen. He lived under few restraints and was equally careless of man, God and nature. Yet he stood by his own standards firmly, remaining brave in battle, loyal to his mates, generous to the Turks, and pledged to his King and country. His speech betrayed few of his enthusiasms, and he accepted success and failure without demonstration, but the confident dash of the horsemen combined with the practical resource and equanimity of the bushman in him, and moved him alike over the wilderness of Sinai and the hills of the Holy Land. Probably his kind will not be seen again, for the conditions of war and peace and romance that produced him have almost entirely disappeared.[31]

Extract from "Remounts"

The breed that's earned its prestige, as a
    "sort that wont wear out,"
Is named "Australian Waler" — guaranteed
    though starved and low —
I'll raise my "felt" to him because, "We
    men who ride them know!"

Trooper Jim Scrymgeour
2nd ALH [1]

Training Hard

Reorganising

Towards the end of the Great War the Citizens Force in Queensland was reorganised; the names and numbers of units were reallocated to perpetuate the traditions and honours of those units which fought during the war. On 17 August 1918 units of the Queensland Mounted Infantry were named the 2nd, 5th, 11th and 14th Light Horse Regiments (Queensland Mounted Infantry). Together they formed the 1st Cavalry Brigade, whose first commander was Brigadier R.M. Stodart who had been the 2nd Light Horse Regiment's first Commanding Officer in 1914.

Shortly after the conclusion of the war in 1918, compulsory military service was introduced in Australia. Naturally the men in the country districts flocked to join the Light Horse units which had fought so valiantly at Gallipoli, and in Sinai and Palestine. The regiments were officered by men, for the most part, who had seen active service and who were prepared to devote their time to train the young lighthorsemen. Regimental Adjutants were Staff Corps officers from the Duntroon Military College and the Warrant Officers from the Australian Instructional Corps, attached to regiments, were responsible for the more detailed training. It wasn't long before the enthusiasm and calibre of the troopers produced both officers and NCOs capable of administering the affairs of a regiment.

In 1921 Lieutenant Colonel G.H. Bourne returned to command his old regiment. He had been a worthy commander during the war, greatly respected by the men, and his return was welcome. Bourne was a dedicated, compassionate and courageous leader who set high standards for both his men and himself. For his actions at Romani he was awarded the Distinguished Service Order and mentioned in despatches. At the end of the war he was selected as an aide-de-camp to the Governor General. Following demobilisation in 1919 Bourne returned to his civilian occupation of manager with the Bank of New South Wales serving in Mackay, Tamworth, Christchurch, (New Zealand) and Rockhampton. Bourne relinquished command of the 2nd LH Regiment (QMI) in 1924, but continued an association with the military, commanding the 33rd Infantry Battalion for three years in the late 1920s, and serving as an air raid warden during World War II. He passed away in 1959 at the age of 78 and is buried in the Lutwyche Cemetery.

There is some confusion regarding the 14th LH following World War I. Military Order 388/1918 changed the title of 27th (North Queensland) Light Horse to 14th (North Queensland) Light Horse, but this did not appear to eventuate and for the next three years there was uncertainty as to the unit's title and location. The 1st Military District records in 1921 indicate that the 14th used drill halls at Bowen, Ayr and Townsville, but correspondence from the unit still displayed "27th (North Queensland) Light Horse". To add to the problem the Officers List of 1920 indicated the 14th LH was at Seymour, in Victoria but didn't have any officers! All was resolved in 1921 when 14th LH was allocated a section of the 2nd LH Regiment (QMI) area and the 14th's headquarters was established at Ipswich.

On Sunday 18 March 1928 the 2nd and 14th Light Horse Regiments (QMI) were presented with Guidons at Enoggera, where the Regiments were in camp for their annual training. Guidons were once carried into battle and were the unit's rallying point or headquarters. The bearer of the Guidon was a soldier of outstanding integrity and someone who had displayed great courage in battle. Many acts of gallantry and self-sacrifice have been performed in the defence of Guidons. Patterned on those of the British Army, Guidons are presented to cavalry units in recognition of outstanding deeds and action in war. Rectangular in shape with two rounded swallow tails they are made of crimson silk damask with the regimental title and colour patch in the centre. These are encased in a wreath of wattle leaves under an imperial crown. The Guidons of the 2nd and 14th share the regimental motto – FORWARD. Both Guidons display eleven battle honours, one for the Boer War and ten for World War I. In 1928 the 2nd LH (QMI) received its Guidon from Brigadier General L.C. Wilson and the 14th LH (QMI) Guidon was presented by Colonel P.J. Bailey the then commander of the 1st Cavalry Brigade. In presenting the Guidon Colonel Bailey told the men, "they were receiving the life blood of the war unit".[2]

The 2nd LH (QMI) was renamed 2nd (Moreton) Light Horse (QMI) and the 14th became the 14th (West Moreton) Light Horse (QMI) to reflect their links with the district. As an economy measure compulsory military training was abolished in 1930 and the Citizens Forces reverted to voluntary service. Both Regiments suffered a loss of manpower with the result that the 2nd and the 14th amalgamated to form the 2nd/14th Light Horse Regiment. The units would remain linked until the outbreak of World War II.

As a result of the Light Horse movement, many men received military training over the years. At troop centres regular parades were held, and at regimental camps (held annually) advanced training was conducted. The annual camps were for the most part held at Enoggera military barracks and the rifle range resounded to the rattle of the Vickers machine-gun, Hotchkiss gun and rifle fire. Sabre troops were trained in swordsmanship and many a competition was held to promote enthusiasm. The swordsmanship event included tent pegging and attacking at a gallop a suspended dummy which had a white four inch target pinned to its chest. A winner of the swordsmanship competition, the then Sergeant Bill Reading, can still recall the thrill of the pounding hooves as he galloped towards the "enemy" on the fields at Enoggera.

Prince of Wales Cup

During this period the Light Horse squadrons polished their

skills with sword, rifle and machine-gun in order to compete for one of several prestigious trophies contested for by lighthorsemen throughout Australia. The Queensland Light Horse Troops competed for the Lord Lamington Shield, the Lord Foster Cup, the W.A. Russell Cup, and the Staff Corps Cup. The Harrisville Light Horse Troop, formed in Harrisville in 1929, under Lieutenant Bob Coppin, became the Premier Troop in Queensland by winning the Lord Lamington Shield in 1931, 1932 and 1933. Because of their prowess, the Harrisville Troop was selected to compete against the Premier Troops from the other States for the prize of prizes, the Prince of Wales Cup. The Troop won the Cup in 1932, the first of only three occasions the trophy was awarded to the Regiment.

The Cup is still the most impressive piece of the Royal Australian Armoured Corps silver collection. Its sheer size and beauty is a sight to behold. It was presented to the Australian Light Horse in January 1904 by the Colonel-in-Chief, General, His Royal Highness, Prince of Wales. The Cup was superbly crafted, in London in 1903, from one solid piece of sterling silver. The silver content is 35 pounds, the distance from handle to handle over three feet, and it can hold in excess of twelve gallons. When the Harrisville Troop won the Cup an admirer filled it with champagne at a cost of £75. The Prince of Wales Cup was first competed for in 1906 by troops "running on their own ground". It carried a £50 prize as well as temporary custody of the coveted trophy. In 1910 it was decided that one team from each state should meet on common ground. That year four teams met at the Williamstown Rifle Range, Victoria. The Cup was contested in 1912, but then not again until 1928. Due to the difficulty of moving competing teams from all states to one location it was decided in 1928 that only New South Wales and Queensland would in future compete for the Prince of Wales Cup. The regiments in the other states were allocated the Hutton Cup for competition.

The competition was a practical test for the troops comprising an advance at the gallop over a one and a half mile course, which included three jumps, to positions from which the enemy (represented by targets) was engaged by rifle fire and then with the sword. The course had to be completed in ten minutes and points were awarded for turn out, saddlery, horsemanship, fire control, shooting, and sword action. The competition was suspended during the Second World War, but recommenced in 1952 by which time the units had been mechanised. The principle of the competition post mechanisation remains the same, armoured vehicles advance over obstacles and engage the enemy with weapons mounted on the vehicles. In 1953 the Reconnaissance Troop of the 2nd/14th succeeded in bringing the Cup back to Queensland. The following year the Cup was awarded to the Beenleigh Troop commanded by Lieutenant Clive Robinson.

Lord Foster Cups

When Lord Foster was Governor General of Australia (1920-25) he took a great interest in the Light Horse Regiments and donated two trophies for competition among the Regiments in two zones in Australia. One was allocated to New South Wales and Queensland and the other to the rest of the states for annual competition between machine-gun troops. The first competition occurred in 1925-26. After a series of eliminating contests, one troop was selected to represent its brigade in the Zone Final.

A course of one mile was selected, and two sets of targets were placed at 500 yards from fire positions. These were engaged by fire from the Vickers guns carried on packhorses to the firing position. On the signal to advance, the troop galloped forward with its packhorses, unloaded the guns and went into action against the "enemy". Each of the eight men in the troop had to fire fifty rounds with the Vickers gun. Guns were then broken-down and placed on the packhorses before galloping to the next fire position. A standard time of fourteen minutes was allowed and, as with the Prince of Wales Cup, points were awarded for horses, saddlery, personnel turnout, and leadership. The coveted Lord Forster was won by Queensland lighthorsemen for the first time in 1937 when the Enoggera based troop of the 2/14 Light Horse Regiment, led by Lieutenant J. Coulter, was successful over the Enoggera competition course. The Cup was again won by the Enoggera Troop in 1939 over the same course, the Troop then commanded by Lieutenant C. Williams.

As well as competing for prestigious trophies, lighthorsemen from all over Australia demonstrated their skills at tent-pegging, sabre charges, Gretna Green races, rescue races, wrestling on horseback, and a myriad of mounted sports which were actually part of the wartime fighting skills and the means of saving their lives, or that of their mates. It is doubtful if many of the spectators, who were thrilled by the sight of a galloping lighthorseman leaning low and using his sword to lift the ring from a tent peg in the ground, realised that this was part of battle strategy. It was a means of bringing down tents on the occupants, to aid the capture of an enemy encampment by mounted troops. Gretna Green and rescue races duplicated the rescue of comrades from enemy territory when their horses had been killed or injured. Wrestling on horseback was merely acting-out a tussle with a mounted enemy.

The Light Horse trained for escort duty and took pride in carrying out these public duties with the precision and "spit-and-polish" that characterises a well-drilled troop in peacetime. In Brisbane the Troop was called upon to provide many an escort for dignitaries including the Governor, Governor-General and notably for the Duke of Gloucester when he visited Brisbane in 1934. All of this required attention to horses, much cleaning and polishing, and of course extra training. The Brisbane Troop also participated in Military Tattoos at the Brisbane Showground to which the troopers responded uncomplainingly and with pride. A further duty was the attendance, mounted, at all Anzac Day parades in Brisbane, keeping the image of the lighthorseman before the public.

Boonah Troop

"Old Light Horsemen" of the Boonah Troop still (1989) meet regularly and the tales they tell of peacetime activities gives us an insight into the fun and excitement (and hard work) of the troops in Queensland. The exploits of the Boonah and Kalbar Troops have been recorded in *Community Service, Past and Present*, a booklet published by the Boonah District Cultural Centre Committee. The Boonah Troop was one of Queensland's most active and enthusiastic Troops in those years between the wars. Formed in 1921, as a part of the 14th Light Horse Regiment (QMI), the Troop was led in turn by Lieutenants Jack Morrish, H. Ryder, Gordon Greenway, Pat Darvall, and Syd Appleby.

In common with other troops throughout Queensland the Boonah Troop held camps, appeared at sports meetings and shows, paraded, and organised military balls where the local debutantes were presented. But unlike the others the Boonah Troop formed a trick-riding team in 1936, they were called the Boonah Cossacks. Alf Betts was a member of the team and recalls the Cossacks with pleasure. The team performed at district shows and for two years was paid £75 per week to appear for twenty minutes nightly at the Brisbane Exhibition.

The Boonah Old Light Horsemen recall how, at one annual camp, a drill sergeant accused the Boonah boys of being the most useless bunch of so-and-so riders he had ever seen. It's doubtful if the lads remembered a great deal of his instruction, but they certainly recall the lovely sight of him attempting to demonstrate the correct posture. The unlucky drill sergeant hit his horse with his spurs and the animal ducked its head, pig-rooted and dumped the drill sergeant neatly on the ground.

They also recall some of the raw recruits who came to the Regiment. There was Luton White who owned Barnes Auto; he joined HQ Squadron transport unit where he assumed he'd drive a truck. He was however given two horses harnessed to a limber! Col Head remembers the camp well:

> Luton was attached to our Squadron while we did a bit of work around the back of Southport and Nerang in mosquito country. Around lunch time, you'd hear the pounding of hooves and rattle of wheels, and the order would be passed up . . . "line the road . . . here comes Lute!" When the outfit came into sight, you'd see Luton hanging on for dear life. He was a motor-man from way back.[3]

Meals were brought on the limbers to the various troops. The food was kept in tightly sealed hot-boxes, which kept the meals hot until delivered. The limbers belonged to HQ Squadron and it amused Col Head and his country mates to discover that all of the limbers were manned by Brisbanites, with little or no experience of horses.

Many indeed were the escapades of the Boonah Troop: golden syrup poured into a Regimental Sergeant Major's (RSM) bed to pay-back an insult; a lizard carried all day in a trooper's pocket so that it could be slipped between the RSM's sheets that night; a dunking in the horse trough for a Lieutenant — he had refused to allow the "boys" the watermelons given to them by the Phillips family of Wyalong homestead.

Most of the fun was harmless such as the time following one Beaudesert camp when the Boonah Troop, with the rest of the Regiment, was on a route march and recruiting drive through Brisbane and up the Brisbane Valley. At a camp near Esk the site happened to be on Mount Brisbane Station — owned by the Esk Troop Leader, Keith McConnell. Keith invited the Commanding Officer and Adjutant to the Mt Brisbane homestead for drinks, and while the hospitality was being enjoyed, one of Keith's stockmen put Keith's tired horse into the horse paddock and replaced it with a fresh mount. The next day the Regiment rode to Ipswich, carrying out a mock battle through Marburg, completing the exercise at Rosewood. The Boonah Troop together with the other troops broke off and made their separate ways home. Before the horses left army jurisdiction however, they had to be inspected — and it's understood that army records still carry no explanation for the fact that Keith McConnell started the exercise with a gelding and ended it with a mare!

All camps were not good fun. On a route march to Southport in 1940, the last time the 2nd/14th Light Horse Regiment paraded with their horses prior to mechanisation, rain fell constantly. Water rushed down Mount Tamborine, making the upward journey a nightmare. One packhorse slipped over the side of a narrow mountain track and was lost. Stan Winoski's horse even "went under" while crossing a flooded creek, leaving Stan bobbing up and down in the water with only his bottom showing. Fortunately he was washed against Les Love's horse and Les grabbed him by the bandolier and dragged him to the bank. The Adjutant, Captain Bill Reading, remembers the nightmarish trip recalling that he spent most of the upward journey on foot hanging onto his mare's tail as she dragged him up the mountain.

While the Boonah boys remember the humorous aspects of their training they appreciated the serious side as well. Although few were to serve on horseback in the coming war, the techniques of weapon handling, fieldcraft, tactics, and discipline acquired during the camps of the 1930s ensured the troopers were prepared to handle the rigours of combat whether they were fighting as infantry or as mechanised soldiers.

Outbreak of War

With the outbreak of war in Europe and Australia's declaration of war against Germany, the government set about mobilising the defence forces. Early in 1940, a cavalry brigade camp was held at Beaudesert, when all three Queensland Light Horse regiments came together. The Brigade conducted regimental level training, tactical exercises without troops (TEWTS) for the officers and held a brigade sports meeting. A recruiting depot was established to enlist volunteers for the Second Australian Imperial Force (2nd AIF). Many men from the regiments enlisted in the AIF for service anywhere in the world in defence of Australia.

A large number of the men who enlisted at Beaudesert were posted to A and B Squadrons of the 2nd/7th Divisional Cavalry Regiment (2/7 Div Cav), then being formed at Redbank AIF camp. The Officer Commanding A Squadron was Major C.V. Boyd, while that of B Squadron was Major W.G. Reading. Both of these officers were from the 2/14 Light Horse Regiment. The two Squadrons, being composed of volunteers almost entirely from the ranks of the 1st Cavalry Brigade of Queensland, provided a sound base for a very efficient unit.

The 2/7 Div Cav equipped with light tanks and machine-gun carriers served in the Middle East until 1942 when, because of the growing threat to Australia by the Japanese, the Division was recalled home. The 2/7 Div Cav Regiment was equipped and trained to fight as an infantry battalion. The modern lighthorsemen recalled that their predecessors had fought dismounted at Gallipoli and now, some thirty years later, lighthorsemen were once again leaving their mounts (armoured vehicles) behind. After landing at Port Moresby, the 7th Division was involved in bitter fighting against the Japanese, the 7 Div Cav Regiment sustaining heavy casualties (including the CO who was killed) at Sanananda. The Division was withdrawn from the frontline, returned to Australia and stationed at Ravenshoe until disbanded.

Exit Horses

The faithful animals which for almost eighty years had carried the lighthorsemen into battle were now to be put out to pasture. It was a sad time for the men, but they realised their four-legged companions could not survive on the battlefield of the 1940s. Mechanisation of the Light Horse, which had begun prior to the war, was now complete. The Queensland based 1st Cavalry Brigade became a mechanised brigade.

In August 1940 the 2/14 LH Regiment was unlinked and the 14th was raised in Eastern Command as the 14th Light Horse (Machine Gun) Regiment. It was renamed the 14th Motor Regiment in March 1942 and in May 1942 was disbanded. The soldiers were sent to the 2/4 Armoured Regiment.

The 2nd Moreton Light Horse (QMI) became the 2nd Reconnaissance Battalion Moreton Light Horse (QMI) in December 1941. It entered camp at Grovely in Brisbane for full-time service. All mounted equipment was withdrawn and replaced with webbing and steel helmets, while tracked carriers were substituted for the horses. In September 1942 the unit became the 2nd Australian Divisional Cavalry Regiment, Moreton Light Horse (QMI) and then in November 1942 the 2nd Australian Cavalry Regiment (AIF). The Regiment was disbanded at Townsville on 29 July 1943. The troops were transferred as reinforcements to the units committed in the New Guinea campaigns.

The Darkest Hours

In February 1942 the Australian mainland was attacked for the first time when Japanese aircraft bombed Darwin in two raids that sank 8 ships, damaged 15 others, destroyed 23 aircraft and caused 545 casualties. Twelve days later, an air raid on Broome destroyed 24 aircraft and caused another 70 casualties. In less than a fortnight the Japanese had caused over 600 casualties. The damage to Australian morale was immense. For the first time the home defences had been breached and the enemy was still advancing south through the island chain; Australians realised that a major crisis was close at hand.

There was barbed wire on the beaches, searchlights and anti-aircraft guns, sandbagged public buildings, "blackouts", slit trenches in the backyard, and the "Brisbane Line" proposal. The Australian and American troops were concentrated near the most populous and industrially vital regions of the southeast corner of Australia, leaving the north lightly defended. Meanwhile Darwin, Broome, Wyndham, and Derby continued to suffering bombing raids which many believed to be a "softening up" process prior to Japanese landings.

In March 1942 a conference involving the Australian Army, Navy and Air force, the United States Army and the Flying Doctor Service of Australia recommended to the government that a northern observer organisation be established to communicate all information of enemy aircraft, naval, and military movements. With urgency the North Australia Observer Unit was to be established to cover the area from Normanton in Queensland to Alice Springs and Derby in Western Australia. To cover the flanks of this unit, observer groups were required to conduct surveillance in Western Australia and North Queensland.

York Force

In May 1942, there was a call for volunteers to man a fully operational AIF Light Horse squadron for service at an as yet undisclosed location. Selection would be made from men with the following qualifications: medically fit, bush experience, horsemanship, initiative, reserve, intelligence, and with preference to those with either signal or intelligence training. For the lighthorsemen who wanted to return to their horses, here was an opportunity that could not be resisted and almost every trooper of the 2nd Reconnaissance Battalion Moreton Light Horse (QMI) volunteered.

The unit, which the men had willingly joined, was the 1st Australian Light Horse Squadron (York Force as it became known). The Squadron was tasked to conduct reconnaissance of York Peninsula, provide warning of Japanese shipping, alert Army Headquarters of any Japanese landing, and then monitor the enemy army's movement once it was ashore. It was believed that Japanese reconnaissance units may have already landed and the enemy's knowledge of North Queensland could be superior to that of the Australian Army. Although an enemy landing on the north coast of Australia, with its mangrove swamps, large tides, and lack of roads would be difficult, the Japanese Army had proven, in Malaya and Papua New Guinea, that no route could be considered impassable.

Initially York Force consisted of a headquarters, four troops (each of two sabre sections), one light machine-gun section, a limber driver, farrier, and so on, a total of 130 men. Several months later the Regiment reinforced the Squadron with an additional sabre section per troop bringing the unit up to full strength. It had been the peacetime practice for lighthorsemen to supply their own mounts, but this time York Force was issued with government remounts. By the time the Squadron deployed on operations there were more than two hundred animals in the horse lines.

Once in location on York Peninsula the need for fresh water for both men and horses was critical, particularly during the dry season when, without water, death was only twenty-four to thirty-six hours away. The Force would bivouac on a creek as near to the mouth as salt water would allow. Once headquarters was established two troops would be despatched in section strength for patrols lasting up to ten days to check old pack tracks over the ranges. The tracks were shown on out-of-date (1927) army maps and, without checking, could not be relied upon. Much valuable information was gained which would have proven vital had a Japanese invasion occurred.

The other two troops would patrol the coastline, each troop moving in opposite directions away from headquarters for a distance of approximately fifty miles and then bivouacing. Sections from these patrolling troops would then conduct dawn-to-dusk coastal patrols with each patrol lasting twelve hours. During the patrols the horses were often required to swim many deep creeks, where, on one occasion, two out of the three man patrol drowned. The threat of enemy contact was always present; observer units in the Northern Territory located enemy radios at Daly River and on Vernon Island, and a number of tracks from Japanese reconnaissance patrols were discovered along the coastline.

Because of the rugged terrain, heavy rainforest and undergrowth, the patrols were very arduous. Vines, stinging trees and thick vegetation made the trips extremely difficult and uncomfortable for both men and horses. A great deal of the terrain was such that in peacetime horses would not have been expected to work there, nor would they have been expected to carry the loads required on patrols, or to work for such long hours without rest. In the wet season, when the work was hardest, feed was at its worst. The very wet conditions caused mould and available fodder was usually inadequate. After two years of continuous operations less than a dozen of the original horses remained.

However, the horses by and large proved the most efficient means of conducting the mission assigned to York Force. In many cases the limber horses of the unit were able to pull trucks and machine-gun carriers from creeks and wet bogs, proving their versatility once again.

On one occasion the Squadron was called out in the middle of the night to dismantle camp and prepare for embarkation. The destination was believed to be India! However there was a last minute cancellation of the movement, much to the men's disappointment. With the halting of the Japanese advance in Papua New Guinea and the clearing of the enemy from the islands to Australia's north by the United States and Australian armies, the threat to Australia lessened. Eventually, the need for York Force decreased and in mid 1944 the Squadron moved to the Black River (north of Townsville) and disbanded. The horses, which had once again served their country well, were turned loose.

Although the 2nd/14th Light Horse did not deploy on active service during World War II, hundreds of its men fought for their country as cavalry and infantry with the 2/7 Div Cav Regiment, as lighthorsemen with York Force, or as riflemen with infantry battalions. The men from the Regiment served as gallantly as did their predecessors in the Boer War and World War I.

> *Let the crowd of men applauding,*
> *When the Light Horse come again,*
> *To mingle Klim Klop with their clapping,*
> *Like flooded creek replete with rain.*
> *Give cheers of Hoop! Hurra!*
> *With the Klim Klop music of their foot-falls*
> *On the wood-paved way.*
>
> P.F. Hockings[1]

Out of the Saddle

Interim Army

After the cessation of hostilities in August 1945 most AIF and militia units began a process of disbandment. Demobilisation and a return to civil vocations were the primary consideration for the greater proportion of the Australian forces. Members of the Regular Army were sorted out as quickly as possible and reposted to appointments in various headquarters and training establishments in Australia. Stores and equipment in good condition and needed for the post-war army were returned to depots for overhauling and storage. The demobilisation plan commenced in September 1945. An "interim" army continued the occupation of former enemy-held territories, to guard prisoners pending their repatriation, handle and care for stores, and to administer the demobilisation scheme. Australia now had a large "built in" reserve of recently demobilised war-time forces and there was no immediate need for a large number of active units.

The end of the "interim" period occurred in late 1947 when Army Headquarters announced the reintroduction of training for the Citizen Military Forces (CMF). The basis of this training, scheduled to commence on 1 July 1948 was for a voluntary enlistment with an initial engagement of two years and subsequently for one year at a time.

Annual training was to occupy a continuous period of fourteen days in camp, with an additional twelve days obligatory and twelve days voluntary home training, that is, at the training depot closest to the Trooper's residence. The Royal Australian

Armoured Corps (RAAC) was asked to raise armoured car, reconnaissance and armoured personnel carrier (APC) regiments. Pre-war Light Horse units totalled twenty-one regiments, but now less than half that number were needed to meet the army's requirement.

Great care was exercised in selecting the units to be raised. Although seniority was important, so also was preservation of the old 1914–18 traditions and territorial links with country districts. But things had changed; this time horses were no longer used by the army and the districts that bred good horses and horsemen were no longer as important to the Regiment. Nor were recruits required to bring along their own horse or truck as had been the case with earlier militia regiments. It was now necessary to look to the larger towns with numerous workshops and possibly organised motor sports to provide suitable recruits. The use of military vehicles for training meant a concentration of personnel in at least squadron strength to achieve an economical management of resources. The horsemen, with their quick eye for country, their capacity to think at a gallop and their traditional dash and initiative were a loss to the Regiment. But in their place came young men equally capable of landing a troop of armoured vehicles, and whose minds were tuned to the speed of the motor vehicle.

The Regiment Reforms

A Squadron 2/14 QMI was formed in 1948 as an independent armoured car unit equipped with Staghound armoured cars. The new armoured unit with a Light Horse name and tradition held a great attraction for the public, and recruiting was strong. Keen recruiting soon brought approval for a full regiment to be raised with the unit organised as follows:

Regimental Headquarters and HQ Squadron	Brisbane
A Squadron	Brisbane
B Squadron	Boonah
C Squadron	Beaudesert

The Regiment set about re-establishing the traditions and customs of the 2nd Light Horse and QMI. Unfortunately it was discovered that one of the unit's treasured possessions was missing. The 2nd LH once owned a magnificent silver and crockery set, a place setting for each officer of the Regiment. Every item, including even the smallest mustard spoon, was embossed with the unit crest. To complement the table settings there were Irish linen tablecloths, embroidered with the regimental badge. The entire collection was housed in large tailor-made wooden chests. Today the 2nd LH silver and crockery would have a value in excess of half a million dollars. As a young officer in the 1930s Bill Reading recalls the settings were complete, although on one occasion he had to retrieve items from a sergeant cook's home. With the threatened invasion of Australia in World War II it is thought the collection was placed in an Ordnance depot for safekeeping, but all efforts since 1948 to locate the mess silver and crockery have proven fruitless. A sad loss indeed!

National Service

National Service was introduced by Parliament in 1951. The terms of the call-up were that at the age of 18 all physically fit young men were expected to serve in the Armed Forces for a period of three years, that is, ninety days full-time duty followed by part-time service in the CMF. The volunteer character of the Regiment was temporarily submerged; existing personnel found themselves at a disadvantage alongside the intensively trained national serviceman. While he generally fulfilled his obligations, the national serviceman failed to show the interest and devotion to the Regiment which is the mark of the militiaman.

In 1956 the length of full-time training for national servicemen was reduced from ninety to seventy days. This was a burden for the Regiment as, the national servicemen were now at a reduced standard of training and the Regiment had to provide them with the experience, as members of an armoured vehicle crew, to ensure their skills were maintained. It had also to introduce them to tactical training. This was no easy task. In fact, it was quite beyond the capabilities of any unit without ample equipment and thoroughly experienced officers and NCOs. The 2/14 QMI, together with other CMF units, found it extremely difficult.

The Band

In 1954 the Regiment raised a band under the conductorship of Captain James Compton. Its first official engagement was to lead the Anzac Day Parade through Brisbane. One of the original members of the band is Sergeant Norm Baxter who began service in World War I, serving overseas with the Light Horse. Norm was to play in the band, until 1971, when after nearly twenty years, failing eyesight forced his retirement.

The band, by popular demand, entered the contesting field. It won a number of trophies, the two most treasured being the certificates for the C Grade Championship of Queensland at Bundaberg in 1969, and the B Grade Championship at Southport in 1970. The latter was perhaps the most meritorious as bad luck struck in the loss of two bass players (due to sickness) and the tenor horn player (departed overseas), just one day before the contest ended. Furthermore, they were in a higher grade than the previous year, against six other bands, one of which was the current holder of both the New South Wales and Australian B Grade Championship. How well they performed is illustrated by the adjudicator's remarks when he told a packed audience that he had never before heard such brilliant playing from a B Grade band!

The most outstanding and memorable duty for the band was in March 1954 when Queen Elizabeth II and the Duke of Edinburgh visited Brisbane. The 2/14 QMI Band was the only CMF band selected to play at a Government House reception in Australia for Her Majesty.

The band performed at various functions for both public and military bodies, such as Queen's Birthday celebrations, Military Tattoos, laying up of Guidons, football matches, and so on. Up until the demise, in 1969, of the South African War Veterans

Association, the band would lead their marches, assist in their ceremonies and were represented at their annual dinner. The President of the Association referred to the Bandmaster, Warrant Officer D.B. Barlow and his band as, "Our adopted boys".

Anti-Tank Regiment

In 1956 the Regiment underwent another change of role. For twenty-five years the anti-tank defence of the army had been the responsibility of artillery, but now the Australian Army, following the British lead, allocated the task to the RAAC. The 2/14 QMI and 3rd/9th South Australian Mounted Rifles were selected to perform this role and the units handed in their armoured vehicles and were issued with 6 and 17 pounder anti-tank guns and Landrovers.

To provide the best anti-tank defence it was often necessary to locate the weapons within the infantry defensive position. The infantry battalions were reluctant to permit the conspicuous 17 pounder inside their wire as it was claimed, with some justification, that these guns would attract immediate counter-measures from the enemy in the form of heavy artillery bombardments. Consequently it was difficult for the 2/14 QMI to conduct comprehensive training with the infantry.

Due to the difficulty in transporting and supporting their tanks away from their main logistic base, the Regular Army (ARA) component of the armoured corps was virtually "locked up" in Puckapunyal, Victoria, thus the 1st Armoured Regiment and their tanks were rarely seen by the rest of the army. Meanwhile the CMF armoured units were experiencing difficulties; they had swollen strengths due to National Service, a shortage of instructors, and inadequate equipment. It was not possible for the CMF to undertake, single-handed, the education of the army in infantry-armoured cooperation on the scale or of the standard necessary. The solution was to disband one squadron of the 1st Armoured Regiment and to use the personnel to raise an ARA squadron in both 2/14 QMI and 4th/19th Prince of Wales Light Horse. The regular squadron was equipped with modified Landrovers and 106 mm RCL (Recoilless Rifle), the remainder of the Regiment continued to use 1 ton trucks, the obsolete 6 pounder and the 120 mm BAT (Battalion Anti-Tank Gun).

The amalgamation provided the ARA personnel with the opportunity to serve away from Puckapunyal and to gain experience in anti-tank training. Underlying the reorganisation was the strongly held view that regular units should be as much a part of the Light Horse tradition as were the Citizen Forces. It was envisaged that 2/14 QMI, and possibly other CMF Regiments, may become ARA in the long term with both ARA and CMF elements continuing a separate existence (but sharing the linked titles and Light Horse traditions). The regular component of the Regiment proved of great benefit to both ARA and CMF soldiers alike.

APC Regiment

The Regiment remained a combined unit until 1960 when the RAAC consolidated into tank, cavalry and APC regiments. This simple, clear-cut arrangement introduced the new cavalry regiment and spelt the end of the anti-tank regiment. The 2/14 QMI was assigned the cavalry role and equipped with Staghound armoured cars, Ferret scout cars and Saracen APCs. At the same time the regular squadron was rena. :ed B Squadron 1st Cavalry Regiment and A Squadron 2/14 QMI reformed as a CMF Squadron.

In 1970 Lieutenant Colonel C.D.F. Wilson assumed command of the Regiment. A Korean War veteran, the Colonel has also commanded the 49th Battalion, Bushman's Rifles. At the annual camp in 1971, at Greenbank, he introduced some infantry minor tactics to the exercise to ensure the troops were proficient in basic bushcraft. The Colonel obviously recalled the Regiment had fought dismounted during the Boer War, at Gallipoli and in Papua New Guinea; thus it was essential the men maintain their basic infantry skills. In 1981 the then Brigadier Wilson returned as Honorary Colonel of the Regiment, an appointment he still holds today. As the President of the Regimental Historical Fund

Committee he has provided guidance and impetus to the development of the Regimental Museum and to ensuring the traditions of the Light Horse are maintained and passed on to the younger members of the Regiment.

Move to Wacol

In September 1971 the Regiment transferred from Moorooka to Sanananda Barracks at Wacol. The area at Moorooka was very small and there had been numerous complaints by neighbouring residents since the introduction of the MII3AI, which was noisier than the Ferret and liable to damage the curbing and road surface. The move to Wacol provided 2/14 QMI with superior facilities and ample space to train. Unfortunately as the Regiment had moved away from its established recruiting base, the move was to have an adverse effect on unit strength.

On 14 November 1971, His Excellency, Governor General of Australia, the Right Honourable Sir Paul Hasluck presented new Guidons to the Regiment. The following year the old Guidons of the 2nd and 14th Light Horse Regiments were laid up in Saint John's Cathedral Brisbane.

In 1972 the annual camp was conducted in the Wide Bay Training Area. It was the first of a number of combined exercises with the ARA 4th Cavalry Regiment (4 Cav Regt). In the coming years such exercises would be held at Wacol, Shoalwater Bay, Charleville, and Somerset Dam. Combined exercises enabled the Regiment to operate with a regular unit, thus providing access to a greater number of vehicles, men and equipment than was normally possible.

The following year the annual camp was held at Singleton, New South Wales. It was the unit's first interstate camp with members travelling by Hercules aircraft to the Dockra airstrip, Singleton. Here, a pool of APCs had been established and the training consisted of reconnaissance and assault troop tactics which culminated in an advance against an "enemy" equipped with anti-tank weapons.

Freedom of the City

On 2 November 1974, the Regiment was granted the Freedom of the City of Brisbane by the Lord Mayor, Alderman Clem Jones. Such an honour is only afforded to persons of dignity or regiments which have rendered conspicuous service and have an association with the city. The 2nd/14th's association with Brisbane had begun in 1860 with the formation of the Brisbane Troop of the Queensland Mounted Rifles. During the 130 year history of the Regiment, the majority of recruitment has been from the Moreton and West Moreton districts — an area within a fifty mile radius of Brisbane. The 1st Contingent to the Boer War had been afforded a Citizens Farewell seventy-five years previously, on 28 October 1899, the organising committee was chaired by the then Mayor of Brisbane, Alderman Seal.

The 1974 ceremony was led by the Commanding Officer, Lieutenant Colonel M.W. Farmer, and consisted of a dismounted parade in King George Square to receive the Scroll and Deed, followed by a mounted parade to exercise the right to enter the city. Prior to the parade, the Regiment contacted the remaining survivors of the Boer War, six of whom attended the ceremony and were honoured by a salute from the Regiment. In his address to the Lord Mayor, the Honorary Colonel, Colonel F.J. Morgan, paid tribute to these veterans:

> Today is a most fitting date for this ceremony, being the 75th anniversary of the embarkation of the Regiment for South Africa. The members on that occasion were also honoured by the citizens of Brisbane at a banquet chaired by the Lord Mayor and at which the Lady Mayoress presented the Regiment with a banner embroidered by the Ladies of Brisbane. (The banner still occupies a place of pride in the Regimental Headquarters at Enoggera Barracks). The traditions established by these volunteers spawned the deeds of Australians in two World Wars and subsequent conflicts. The Regiment is so happy to have Boer War veterans Messrs. L.T. Jackson, A.E. Ingold, B.J. Richardson, J.T. Picton, Ernest Atkins and Archibald King witnessing the ceremony and I am certain all present share this happiness, particularly for Trooper King, the last surviving member of the Queensland Mounted Infantry who saw service in South Africa. The guards' salute to you gentlemen was your fit and

proper entitlement. Your presence here, bearing in mind your ages, is an inspiration to us all and an excellent example of free going spirit for the young members of the Regiment.[2]

Mr Archibald King was 98 years old at the time of the parade. Following his military service he pioneered the pineapple industry in the Glasshouse Mountains area. Mr J.T. Picton had been Baden-Powell's runner during the siege of Mafeking and had been successful in slipping through the Boer lines. During World War I he was an RSM and in World War II a Captain of a US Small Ship.

Also attending were a number of personalities who had been associated with the Regiment over the years. Lieutenant Colonel F.C.E. From had been the Commanding Officer at the outbreak of World War II. He was a former Danish cavalry officer, losing an arm at the battle of the Somme. Notwithstanding his disability, he was renowned for his tent-pegging feats, controlling his horse reins by his teeth. Major Sydney Appleby was an original member of the 14th Light Horse, taking part in the Regiment's last mounted parade at Southport in 1941 and commanding long range patrols into Cape York during 1943. As unit historian he had been instrumental in arranging the Freedom of the City for the Regiment. Mrs Jean McGregor-Lowndes, as the Patroness of the South African War Veterans Association, had a long and active association with the surviving veterans.

Combined Training

During the 1970s the Regiment continued to experience difficulty maintaining its numbers, despite several intensive recruiting campaigns. On 31 July 1976 the unit was reduced to one squadron and retitled A Squadron 2nd/14th QMI, with Major Gerry Stapleton as the Officer Commanding. Shortly afterward the 2/14 QMI Museum Trust was established as a separate entity with Captain Graham Jardine-Vidgen as curator. He prepared an elaborate and detailed display for the museum, which was originally housed in the conference room at Wacol but eventually a separate building was obtained as more exhibits were donated. The museum is in existence today at the Regimental Headquarters and houses medals, photographs, diaries, letters, uniforms, weapons, vehicles, and so forth, collected during the unit's long and active history.

In 1979 in a combined exercise with 4 Cav Regt, a rare experience for the younger soldiers included travel on a troop train from Albion to Gympie and a road run with over one hundred armoured vehicles from Gympie to the Wide Bay training area. The main element of the exercise consisted of a three squadron advance from Wide Bay towards Brisbane. During the exercise a Public Relations display was established for the residents of Kilkivan.

Later that year it was decided that the Regiment would adopt the wearing of a Stable Belt, then being introduced throughout the army. The colours of blue over green were selected – the colours of the 2nd and 14th Light Horse Regiments. The green of the 2nd, which is the senior Regiment, is above the blue of the 14th.

Proposal to Unlink the Regiments

During 1979 the Commanding Officers of 2/14 QMI and 4 Cav Regt discussed the feasibility of unlinking the 2nd/14th, with the Army Reserve retaining the title 2nd Light Horse while 4 Cav Regt would be designated 14th Light Horse. Such a move would have benefitted 4 Cav Regt, which at the time had only been in existence for four years and had no approved unit badge, Guidon, battle honours or traditions. It would have enabled the traditions of the Light Horse to be shared with the Regular Army and ensure that two famous Light Horse Regiments were retained on the Order of Battle. A detailed proposal to unlink 2/14 QMI was prepared by Lieutenant Colonel R.M. Earle, the Commanding Officer of 4 Cav Regt. Strong support was received from the Commander 1st Division, former Regimental CO's and former Honorary Colonels. Unfortunately the submission fell on deaf ears and the unlinking of 2/14 QMI was rejected by Army Office in Canberra. This prompted the Honorary Colonel to approach the Chief of the General Staff (C.G.S.) In his reply, declining the proposal, the C.G.S. stated:

> I am sure you will agree that Regular Army units unlike units of the Army Reserve which are territorially based, are subject to changes in location from time to time for a number of reasons, such as economy and efficiency. To preserve this element of flexibility and response on a national basis, it is simple and logical to maintain a numerical and functional system in the designation of Regular Army units. It was in this context that the proposal as a whole was not approved. Were I to accede to this proposal I would be introducing a system which in the course of time would lead to confusion as Regular Army units bearing particular regional titles, would move to regions unconnected and unconcerned with such titles.[3]

Strangely this policy, issued in 1980, had little bearing on events only six years later.

Beenleigh

In 1979 it was decided to raise an element at Beenleigh, using the local high school as a venue for parades. By December that year 12 recruits were regularly paraded at the school; thus it was decided to raise a troop around this nucleus. Eventually the Beenleigh Depot raised and maintained a troop of thirty personnel. By mid-1980 the Squadron's overall strength had risen to 232, excluding the Band, and with a very commendable retention rate of eighty-three per cent. On 3 December 1980, 2nd/14th QMI was once again a Regiment!

A proposal was put forward to change the title from 2nd/14th QMI to 2nd/14th Light Horse Regiment QMI, to more accurately reflect the history and traditions of the Regiment. The proposal was approved in 1982 but as the change did not include the word "Regiment", the title became 2nd/14th Light Horse (QMI), the title the Regiment carries today.

At the Commonwealth Games, held in Brisbane in 1982, the Regiment was awarded the honour of providing a guard (command by Major Dick Magnus) at the handing over of the Commonwealth Games flag to the Lord Mayor of Brisbane. Once again the long association between the unit, the city and its people was affirmed.

Integration

In 1986 the Regiment began a new chapter when it integrated with the ARA 3rd/4th Cavalry Regiment. The title of 3rd/4th Cavalry Regiment was transferred to B Squadron in Townsville and the 2nd/14th Light Horse (QMI) became an integrated Regular Army and Army Reserve unit. An Integration Parade was held at Enoggera Barracks on Sunday 29 June, 1986 and reviewed by Major General K.G. Cooke, AO, RFD, ED. In his address to the soldiers he said it was the first time that two major units, one manned by Regulars and one manned by Reservists, had been amalgamated to form a single integrated unit. General Cooke welcomed the integration as the means to save the 2/14 LH and to prevent 3/4 Cavalry Regiment from being reduced to squadron strength.

> Now not only will both be saved, but (as a combined unit) will have a new and challenging lease of life. By combining the two units we will still be able to maintain an APC Regiment in the Australian Army and the new-look 2/14 Light Horse will be an essential, high priority and on-going role. It will be the Army's sole repository of APC skills at Regimental level. By coincidence, or maybe not so much coincidence, I believe this could not have happened in a better way. After all, the original forebears of 3/4 Cavalry Regiment did come from Army Reserve armoured units. Indeed, 4 Cav was a direct product of 2/14 Light Horse Regiment in the 1960s. Now we are putting the pieces of this genealogical jigsaw back together again, to create a unit with a history and traditions stretching back over 125 years to 1860, which included distinguished service stretching from South Africa in 1899, covering the famous World War I battles at Gallipoli, and in the desert campaigns.[4]

Within months of integration the Regiment was thrust into its first major exercise as a fully integrated unit. Exercise "Diamond Dollar 86" involved the Regiment providing support to the regular infantry battalions of the 6th Brigade at the Shoalwater

Bay Training Area. The exercise used conventional warfare tactics involving the advance and defence phases of war. It culminated with the live firing of a variety of weapons including artillery, aircraft, mortars, rifles, and APC mounted machine-guns. For some soldiers it was their first opportunity to train with the Regular Army and to fire their weapons under "battle" conditions. All of the men performed well, living up to the proud traditions of the Regiment, and resulting in 2/14 Light Horse Regiment (QMI) being commended for its effort.

Since integration the Regiment has supported the regular and reserve infantry battalions and in 1989 participation in two major field exercises. The first at Shoalwater Bay Training Area comprised the Regiment opposing a United States Marine Corps amphibious landing, providing the soldiers with the opportunity of observing and competing against one of the best-equipped forces in the world. Late in 1989 the 2/14 Light Horse (QMI) participated in exercise "Kangaroo 89". This is the army's largest exercise and is held every two or three years. The exercise involved the 1st Division, supporting troops, Australian Navy, Australian Air Force and units from overseas forces. "Kangaroo 89" was conducted in northwestern Australia near Kunannurra on the Western Australian/Northern Territory border. The Regiment flew to the exercise area, the vehicles were transported by road and rail. The 2/14 QMI was involved in low-level operations against an insurgency enemy. Once again the troopers performed in a credible manner displaying to the rest of the Australian armed forces the professionalism for which the Light Horse is renowned.

For 130 years the Regiment has been an essential element of Queensland's and Australia's defences. The Queensland Mounted Infantry's reputation was established in the Boer War during fighting at Sunnyside, Kimberley and Elands River. During the World War I this distinguished record was enhanced by the 2nd Light Horse when charging at Quinn's Post, fighting a rearguard action at Romani, defending the Ghoraniye Bridgehead, and holding on at all costs at Abu Tellul; and by the 14th Light Horse at Musallabeh and during operations in Palestine and Syria. In World War II men of the Regiment once again fought in defence of their country as members of the 7th Divisional Cavalry Regiment in the Middle East and Papua New Guinea, as lighthorsemen with York Force, and as reinforcements with infantry and armoured units.

During the life of the Regiment there have been many changes: alterations to the unit's title, conversion from mounted infantry to lighthorsemen and finally mechanisation, and the integration of reserve and regular troopers. No doubt the saddest event occurred in the 1940s when the horses were replaced by armoured vehicles. Through the years the officers and men have readily adapted to the changing events proving the lighthorsemen have heeded the unit's own motto... FORWARD.

Epilogue

On Sunday 10 July 1988, Mr Gordon Williams, the last of the Queensland Mounted Infantrymen, passed away at the Greenslopes Repatriation Hospital, aged 108 years. He was born on Bendina Station near Cunnamulla and spent most of his life on cattle stations, retiring at the age of 79. Gordon attributed his longevity to "his young days riding horses and his daily tot of rum".[5] He served with the 7th Australian Commonwealth Horse in South Africa in 1902 under the command of Lieutenant Colonel Chauvel. During World War I he served in the Middle East with the 2nd Light Horse Regiment.

His funeral was attended by serving members of 2nd/14th Light Horse (QMI) who played "The Last Post" and fired a volley to farewell the last lighthorseman.

Appendixes

1. Honour Roll of the Queensland Mounted Infantry
2. Honour Roll of the 2nd Light Horse Regiment
3. Honour Roll of the 14th Light Horse Regiment
4. Battle Honours
5. Awards for Gallantry
6. Order of Battle
7. Commanding Officers

Appendix 1

Honour Roll

Queensland Mounted Infantry Contingents
Boer War

The following made the supreme sacrifice:

Contingent	Number	Rank	Name	Date	Remarks
First	134	Private	Broderick, E.ST J.V.	25.6.00	Enteric
First	95	Private	Bryce, W.H.	24.8.01	Died of wounds while serving with Prince of Wales L.H. as Sergeant Major
First	56	Corporal	Conley, G.B.	31.3.00	Killed in action
First	180	Private	Cumner, T.	20.3.00	Enteric
First	242	Private	Damrow, W.A.	19.12.00	Enteric
First	177	Bugler	Devitt, W.	12.5.00	Enteric
First	219	Private	Jones, V.S.	1.1.00	Killed in action
First	91	Private	McLeod, D.C.	1.1.00	Killed in action
First	165	Private	Reece, H.L.	31.3.00	Killed in action
First	25	Private	Walker, S.	25.6.00	Enteric

Appendix 1

Contingent	Number	Rank	Name	Date	Remarks
First	116	Corporal	Wriford, G.	11.3.01	Cause not known, whilst serving with Provincial Mounted Police
Second	108	Private	Cronan, E.	2.4.00	Enteric
Second	128	Private	Landsborough, S.L.	15.6.00	Pneumonia
Second	90	Private	Parnell, J.	21.7.00	Pneumonia
Second	54	Private	Reimers, C.M.	2.4.00	Enteric
Third		Lieutenant	Annat, J.W.	6.8.00	Killed in action
Third	157	Private	Bailey, A.T.	19.1.01	Enteric
Third	294	Private	Hull, G.	6.1.01	Enteric
Third		Lieutenant	Leask, J.	20.8.00	Died of wounds
Third	28	Private	MacDonald, D.K.	12.2.01	T.B.
Third	31	Private	Maude, A.E.F.C.	20.8.00	Pneumonia
Third	40	Private	Masterton, J.T.	11.9.00	Died of wounds

Taken from "Official Records of the Australian Military Contingents to the War in South Africa" by Lieutenant Colonel P.J. Murray R.A.A.

Appendix 2

Honour Roll

2nd Australian Light Horse Regiment (AIF)

The Great War

During the War there passed through
the 2nd Australian Light Horse Regiment (AIF)

Officers 103
Other Ranks 2508

Summary of Casualties

Killed 201
Wounded 458

The following made the supreme sacrifice:

Number	Rank	Name	Date
292	Trooper	Arthur, L.R.	7.8.15
293	Trooper	Alexander, C.B.	7.4.15
459	Corporal	Alexander, J.C.	14.5.15
596	Trooper	Anderson, A.E.	15.5.15
611	Trooper	Adams, W.H.	14.5.15
755	Trooper	Anderson, A.	7.8.15
927	Trooper	Allen, F.C.	11.11.15
962	Lance Corporal	Anthony, A.H.	9.1.17
1080	Trooper	Angus, M.	9.1.17
1229	Trooper	Ansell, F.	31.10.17

Appendix 2

Number	Rank	Name	Date
1307	Trooper	Archibald, R.	3.2.16
1537	Trooper	Arthur, C.	24.11.17
2928	Trooper	Archibald, L.L.	11.4.18
	Lieutenant	Burge, J.	7.8.15
96	Sergeant	Barry, H.J.	7.8.15
124	Sergeant	Barnes, J.P.	9.1.17
299	Sergeant	Brown, T.M.	20.4.17
472	Trooper	Burton, A.A.	14.5.15
483	Trooper	Beyers, J.A.	14.5.15
485	Sergeant	Bond, C.J.	30.5.15
592	Trooper	Buchanan, C.G.G.	14.5.15
723	Trooper	Butler, E.R.	14.5.15
1083	Trooper	Boyle, J.J.	12.1.17
1543	Trooper	Bailey, R.	9.1.17
1548	Trooper	Brown, F.	28.3.18
2302	Trooper	Bryce, W.	3.12.17
2868	Trooper	Bunkum, W.E.H.	14.7.18
3349	Trooper	Beck, C.	31.10.17
375	Trooper	Brown, H.S.	28.5.15
	Major	Chambers, A.F.	20.4.17
32	Driver	Carl, W.	7.8.15
134	Sergeant	Cowie, J.	14.7.18
135	Sergeant	Cowley, E.C.	9.1.17
273	Squadron Quartermaster Sergeant	Cartwright, F.	18.4.15
311	Farrier Sergeant	Chambers, J.A.	16.7.18
489	Trooper	Crowther, S.	14.5.15
490	Driver	Casey, P.	14.5.15
797	Trooper	Campbell, D.D.	4.8.16
859	Lance Corporal	Cameron, C.	9.1.17
1400	Trooper	Cunningham, T.A.	4.8.16
1316	Trooper	Cummins, G.	7.11.17
2247	Trooper	Coney, W.	14.7.18
2634	Trooper	Crozier, J.F.	11.1.17
2635	Trooper	Crozier, R.J.	11.4.18
2872	Trooper	Callaghan, P.	19.4.17
3114	Trooper	Cooper, R.	27.3.18
3175	Trooper	Cornick, G.F.	14.7.18
3414	Trooper	Collett, P.P.	11.4.18

Number	Rank	Name	Date
3471	Trooper	Cain, M.W.	18.10.18
3479A	Trooper	de Crespigny, P.C.	14.7.18
745	Corporal	Cochrane, J.W.	20.7.15
142	Trooper	Day, A.	11.2.17 (While Prisoner of War)
312	Trooper	Demmick, L.	3.3.16
461	Sergeant	Drysdale, G.R.	11.2.17 (While Prisoner of War)
504	Trooper	Davies, T.	14.5.15
618	Trooper	Denton, V.	31.5.15
735	Trooper	Durham, G.T.	31.10.17
741	Trooper	Dwyer, J.T.	7.8.15
1155	Trooper	Davies, W.J.	22.12.17
1241	Trooper	De Raadt, J.L.C.	17.10.15
1304	Trooper	Dinsdale, T.S.	1.11.17
2165	Trooper	Daniels, W.E.	11.4.18
2329	Trooper	Dale, T.	4.8.16
2994	Trooper	Deberg, A.H.	14.7.18
317	Lance Corporal	Easton, F.P.	19.11.16 (While Prisoner of War)
646	Corporal	Emerson, J.E.R.	11.4.18
805	Sergeant	Ellis, H.V.V.	22.10.18
1160	Trooper	Elliott, W.	17.7.18
150	Trumpeter	Foote, N.V.	7.4.15
311A	Sergeant	Floyd, L.L.	8.3.18
601	Trooper	Faircloth, B.	14.7.18
1018	Trooper	Finnis, P.W.	4.8.16
2877	Corporal	Fisher, C.H.	9.1.17
	Major	Graham, D.M.L.	14.5.15
155	Sergeant	Geddes, (M.M.), J.R.	14.7.18
328	Trooper	Geoghengan, A.M.C.	15.9.15
462	Corporal	Graffunder, A.C.J.	14.5.15
514	Trooper	Garvey, L.P.	14.5.15
517	Trooper	Goodall, T.L.	14.5.15

Number	Rank	Name	Date
772	Trooper	Gibbs, S.J.	21.7.16
1091	Sergeant	Grau, F.W.	14.7.18
1421	Trooper	Gwynne, T.K.	2.11.18
3239	Trooper	Grieve, C.J.	16.7.18
	Captain	Handley, W.J.	16.7.18
	Lieutenant	Hinton, H.G.	7.8.15
160	Trooper	Harman, H.	2.8.15
523	Trooper	Hannah, J.	14.5.15
529	Trooper	Hulbert, P.R.	18.7.15
624	Trooper	Harris, O.	17.5.15
771	Trooper	Hammond, H.	7.8.15
869	Trooper	Hamp, S.S.	7.8.15
1003	Trooper	Harte, M.A.	3.1.16
1094	Corporal	Hanslow, N.	22.4.17
2809A	Trooper	Haigh, C.W.	6.11.18
2999	Trooper	Hildebrand, T.A.	14.7.18
3240	Trooper	Irish, V.C.	14.7.18
169	Trooper	Jury, M.V.	25.8.15
1029	Trooper	Jones, T.	6.12.15
	Lieutenant	Kemp, A.C.	14.4.18
	Lieutenant	King, (M.C.), W.K.	14.7.18
170	Trooper	Keid, W.	23.6.15
340	Trooper	Kerr, J.	29.6.15
341	Trooper	Kimber, J.H.	7.5.15
342	Trooper	Kelly, J.A.	17.5.16
1227	Trooper	Kirk, A.C.	11.4.18
1248	Trooper	Kelsall, J.	4.11.17
2146	Trooper	Kennett, V.I.N.	9.2.17 (While Prisoner of War)
	Major	Logan, T.	7.8.15
	Lieutenant	Letch, H.A.	22.8.18 (With Flying Corps)
103	Sergeant	Linnan, P.P.	4.11.18
247	Trooper	Lush, J.R.	7.8.15
780	Corporal	Langridge, T.E.	18.5.17
882	Trooper	Lord, A.C.	6.12.15

Number	Rank	Name	Date
1100	Corporal	Larkin, S.C.	28.10.18
3300	Trooper	Loman, T.A.	30.7.18
	Major	Markwell, (D.S.O.), W.E.	31.10.17
17	Trooper	Mouritz, L.B.	14.5.15
21	Driver	McDonald, A.	2.11.17
74	Driver	McCreedy, R.	16.11.15
176	Signaller	McAllister, W.J.A.	24.5.15
179	Trooper	McGowan, E.	22.5.15
180	Trooper	McNamara, L.	20.5.15
234	Trooper	Masters, G.A.L.	5.7.15
241	Trooper	McMahon, J.P.	15.3.16
347	Trooper	Marson, C.	7.8.15
348	Trooper	Moran, W.G.	29.6.15
352	Sergeant	Matthews, A.	9.1.17
356	Corporal	Masters, S.G.	14.7.18
358	Trooper	MacFarlane, C.J.	3.7.15
538	Trooper	McIndoe, E.J.	5.6.15
541	Lance Corporal	Mulvey, F.C.	14.5.15
545	Trooper	Martin, G.B.	1.11.17
600	Trooper	Murray, D.	17.6.15
786	Trooper	McKinnon, D.	15.12.15
787	Trooper	Milford, W.	11.8.15
888	Trooper	McMartin, A.G.	1.11.17
952	Trooper	Marks, W.L.	11.4.18
1105	Trooper	McLean, H.A.	3.12.15
1182	Trooper	Morgan, J.A.	14.8.16
1628	Trooper	McCann, W.E.	11.4.16
	Major	Nash, A.W.	29.6.15
366	Trooper	Norton, W.T.	19.6.15
1721	Corporal	Nash, I.T.	5.3.16
370	Trooper	O'Connor, W.H.	19.6.15
423	Trooper	Ogilvy, J.	2.8.15
557	Sergeant	Oswin, A.E.	14.5.15
945	Trooper	O'Leary, J.	11.11.15
2477	Trooper	O'Callaghan, T.	4.8.16
3287	Corporal	Ogg, T.A.	14.7.18
3250	Trooper	O'Connor, L.	16.7.18
95	Farrier Sergeant	Pickering, A.	10.12.17
190	Trooper	Pearce, E.C.W.	7.8.15

Number	Rank	Name	Date
198	Trooper	Parker, K.A.	20.4.15
199	Trooper	Phillips, T.H.	14.5.15
719	Trooper	Perrott, H.J.	12.10.15
1111	Trooper	Peterson, E.	3.11.15
1261	Trooper	Perkins, T.O.	11.4.18
1341	Trooper	Parkes, W.J.	9.8.16
3491	Trooper	Peach, W.J.	14.7.18
201	Trooper	Quinn, J.W.	22.12.15
	Lieutenant	Righetti, A.S.	4.8.16
378	Trooper	Robertson, G.H.	13.3.15
382	Trooper	Ross, J.R.	9.9.15
383	Trooper	Reade, C.M.	30.5.15
1688	Trooper	Robertson, F.J.K.	9.1.17
	Lieutenant	Sinclair, D.H.M.	20.4.17
	Lieutenant	Swanston, W.	1.11.17
33	Driver	Sefton, R.H.	11.11.15
214	Squadron Quartermaster Sergeant	Stewart, H.J.	3.9.15
388	Trooper	Sinclair, H.	15.10.18
593	Trooper	Sims, J.C.	14.5.15
713	Trooper	Stower, R.W.	31.10.17
794	Trooper	Smale, W.E.	7.8.15
795	Trooper	Sharpe, W.	13.8.15
891	Trooper	Stevenson, T.	26.9.15
1046	Trooper	Spry, M.	16.11.15
1266	Lance Corporal	Somerville, J.	10.4.17 (While Prisoner of War)
1732	Trooper	Sullivan, G.	4.8.16
2218	Trooper	Sullivan, J.	1.11.18
2388	Trooper	Sweedman, E.C.	15.7.18
4052A	Trooper	Sullings, H.A.	31.10.17
570	Trooper	Templer, F.D.	22.10.15
571	Trooper	Tallentire, W.J.	14.5.15
1049	Trooper	Toohey, W.	4.8.16
2381	Lance Corporal	Underwood, T.	22.1.17
645	Trooper	Underhill, R.M.	14.5.15
	Lieutenant	Woodyatt, P.S.R.	4.8.16

Number	Rank	Name	Date
230	Lance Corporal	Wilson, G.L.	19.5.15
236	Driver	Whittall, P.G.	7.8.15
399	Corporal	Waddell, G.M.	23.12.16
456	Sergeant	Wade, J.S.	14.5.15
584	Trooper	Wentford, J.F.J.	21.5.15
586	Trooper	Wilson, A.G.M.	14.5.15
647	Trooper	Wragge, C.E.	16.5.15
803	Trooper	Wilson, W.T.	7.8.15
804	Trooper	Wylie, E.	9.8.15
973	Trooper	Ward, J.E.N.	6.3.17 (While Prisoner of War)
1117	Trooper	Wallace, A.G.	31.10.17
1121	Trooper	Watt, A.J.	23.12.15
1445	Trooper	Weeks, G.R.	31.10.17
1472	Driver	Wooster, A.C.	2.11.17

Taken from Appendix to "The History of the 2nd Light Horse Regiment (Australian Imperial Force) August 1914 – April 1919" by Lieutenant Colonel G.H. Bourne, D.S.O.

Appendix 3

Honour Roll

14th Australian Light Horse Regiment (AIF)

The Great War

The following made the supreme sacrifice:

Number	Rank	Name	Date
2245	Private	Antonson, H.L.	16.10.18
R27	Warrant Officer	Aughey, J.G.	30.9.18
2888	Private	Cooper, J.	7.11.18
	Lieutenant	Cox, J.P.	19.9.18
3490	Private	Eddy, D.H.	12.10.18
2762	Private	Farrell, J.S.	11.10.18
	Major	Hogue, O.	3.3.19
2125	Private	Keshan, F.A.	11.12.18
2822	Private	Mitchell, C.R.	29.8.18
1109	Private	Vigars, H.J.	30.9.18
2231	Corporal	Walter, W.J.	20.3.19

Appendix 4

Battle Honours

The 2nd/14th Light Horse (Queensland Mounted Infantry) carries the Guidons of 2nd Light Horse Regiment (Queensland Mounted Infantry) – Moreton Light Horse, and 14th Light Horse Regiment (Queensland Mounted Infantry) – West Moreton Light Horse.

Battle Honours for the Boer War were presented to the existing Light Horse units in 1908 in recognition of the representation by the states in the war.

Battle Honours for the Great War (World War I) were the result of service by 2nd Australian Light Horse Regiment (Australian Imperial Force) and 14th Australian Light Horse Regiment (Australian Imperial Force). In the latter case, as the Regiment was formed late in the Great War from the 1st Australian Battalion, 1st Brigade, Imperial Camel Corps, it is understood that the Regiment received the Battle Honours appropriate to that Battalion.

Battle Honours

The 2nd Light Horse Regiment
– Moreton Light Horse (QMI)

The Boer War   —   South Africa, 1899-1902*

The Great War   —   Anzac, Defence of Anzac*, Suvla, Sari Bair*, Gallipoli, Rumani*, Magdhaba-Rafah*, Egypt 1915-17, Gaza-Beersheba*, El Mughar, Nebi Samwil, Jerusalem*, Jaffa*, Jericho*, Jordan (Es Salt), Jordan (Amman)*, Megiddo*, Nablus, Palestine 1917-18.

*Indicates those displayed on the Guidons.

The 14th Light Horse Regiment
– West Moreton Light Horse (QMI)

The Boer War   —   South Africa, 1899-1902*

The Great War   —   Rumani*, Magdhaba-Rafah*, Egypt 1916-17*, Gaza-Beersheba*, El Mughar, Nebi Samwil, Jerusalem*, Jordan (Es Salt)*, Jordan (Amman)*, Megiddo*, Nablus*, Damascus*, Palestine 1917-18.

*Indicates those displayed on the Guidons.

Appendix 5

Awards for Gallantry

The following decorations were awarded to members of the Queensland Mounted Infantry:

Brosnan, J.C.	Corporal	MID
Chauvel, H.G.	Major	CMG
Cooney, T.	Quartermaster Sergeant	DCM
Davidson, N.A.	Corporal	DCM
Duka, A.T.	Captain (Surgeon)	DSO
Forbes, A.E.	Bugler	DCM
Glascow, T.W.	Lieutenant	DSO
Gordon, R.	Captain	DSO
Harris, A.A.	Corporal	DCM
Harris, H.	Corporal	DCM, MID
Keogh, H.W.	Bugler	DCM
Loynes, J.	Sergeant	MID
Reid, D.E.	Captain	DSO
Ricardo, P.R.	Major	CB
Ryan, J.P.	Sergeant	MID
Stevens, W.J.	Corporal	MID
Tancred, G.H.L.	Corporal	MID
Trickett, J.J.	Corporal	DCM
Tunbridge, W.H.	Major	CB
Walker, J.J.	Company Sergeant Major	DCM
Wright, W.L.F.	Quartermaster Sergeant	DCM

The following decorations were awarded to members of the 2nd Light Horse Regiment.

Allen, F.J.	Squadron Sergeant Major	DCM
Apelt, H.A.	Sergeant	MM
Baldwin, A.	Corporal	DCM
Barton, F.J.	Lieutenant	MC*
Biggs, A.F.	Sergeant	MM
Birkbeck, G.	Major	DSO, MID
Bourne, G.H.	Lieutenant Colonel	DSO, MID
Brown, R.B.	Lieutenant	MC*
Brown, W.J.	Major	DSO
Butler, J.H.	Lieutenant	MC*
Cameron, E.C.B.	Lieutenant	MC*
Campbell, A.E.G.	Major	DSO, MC*
Campbell, J.H.	Lieutenant	MM, Croix de Guerre
Cannons, E.J.A.	Sergeant	MID
Carlyon, J.E.	Sergeant	MM, MID
Carroll	Lieutenant	MC*
Chisholm, A.	Major	DSO
Cory, W.M.B.	Lieutenant	MC*
Duff, C.E.	Trooper	MM
Emmerson, J.E.	Corporal	MM
Evans, F.	Captain	MC, MID
Finlayson, T.	Corporal	MM
Frankford, H.L.A.	Warrant Officer	MSM
Franklin, R.N.	Major	DSO
Geddes, J.R.	Sergeant	MM, MID
Glasgow, T.W.	Major General	KCMG, CB, DSO Croix de Guerre*
Glascow, R.	Major	DSO, MC*
Gordon, H.K.	Chaplain	MC, MID
Groffinder, A.C.	Corporal	MC
Guiren, L.B.	Lieutenant	MC
Henderson, L.J.	Lieutenant	MC
Jones, P.C.	Signaller	MM
Kempson, W.H.	Trooper	Act of Gallantry

Keid, W.	Trooper	MID
Kenny, W.H.	Trooper	DCM, Medaille Militaire
King, W.K.	Lieutenant	MC
Kirkbridge, T.	Sergeant	DCM, Cross of Karra George
Leander-Grove,	Lieutenant	MC
Leton, H.A.	Lieutenant	MC, MID
Letch, H.A.	Lieutenant	MC, MID
Locke, J.	Sergeant	MID
McCartney, G.W.	Lieutenant Colonel	DSO*
McDonald, J.M.	Corporal	DCM
McDougal, C.M.	Lieutenant	MC
McDougall, M.D.	Captain	MC
McKnee, E.	Trooper	MID
McSharry, T.	Lieutenant Colonel	CMG, DSO, MC*
Markwell, W.E.	Major	DSO, MID (twice)
Massey, W.H.	Trooper	MM
Menzies, J.A.	Lieutenant	MC*
Menzies, J.A.	Squadron Sergeant Major	MID
Mercer, J.D.	Sergeant	MID
Morgan, G.O.	Sergeant	MID
Mowbray, C.	Sergeant	Croix de Guerre, MID
Nicholson	Lieutenant	MC*
Ogilvy, A.J.	Captain	MID
Oswin, A.E.	Sergeant	MID
Pearson, J.	Sergeant	MID
Pledger, G.T.	Lieutenant	MC
Power, G.W.	Sergeant	MID
Pringle, F.L.	Trooper	DCM
Shanahan, M.	Major	DSO
Shand, M.H.	Trooper	MID
Shire, M.H.	Trooper	MM
Sinton, R.G.	Lieutenant	Serbian Silver Star

Stodart, C.C.	Major	MC, MID
Tranter, C.J.	Trooper	Serbian Gold Medal*
Trinca, F.	Captain	MC
Volp, G.	Corporal	MM
Wade, J.S.	Sergeant	MID
Wasson, J.	Lieutenant	Medaille Militaire, MID
Wilcox, F.	Lieutenant	MC, MID
Willenbroc	Captain	MC
Wills, J.M.	Lieutenant	MM, MID
Wilson, A.J.	Sergeant	MM

*Awarded while seconded from the Regiment or after transfer to other units.

The following decorations were awarded to 1st Battalion Imperial Camel Corps or 14th Light Horse Regiment members:

Callander, G.A.	Sergeant	MSM
Campbell, A.E.G.	Major	DSO, MC, Croix de Guerre
Cashman, W.P.	Lieutenant	MC
Clarke, J.	Corporal	MM
Denson, H.R.	Major	DSO
Dickinson, H.G.	Private	MM
Donaldson, A.	Trooper	MM
Garland, H.W.	Private	MM
Gowland, G.	Lieutenant	MC
Greenway, C.	Squadron Sergeant Major	DCM
Gull, A.W.	Lieutenant	MC
Haseler, R.E.	Corporal	DCM
Hill, A.N.	Sergeant	MM
Holland, J.F.	Lieutenant	MC
Kellow, J.H.	Private	MM

Awards for Gallantry

Langley, G.F.	Lieutenant Colonel	DSO, Serbian Order of the White Eagle, 5th Class (with swords)
Long, H.	Private	DCM
Love, J.R.B.	Lieutenant	MC, DCM
McCallum, S.	Corporal	DCM
McGrath, J.P.	Private	DCM
McGrath, W.F.	Private	MM
Mackenzie, A.A.	Lieutenant	MC
Malcolm, H.L.D.	Lieutenant	MM
Mason, J.B.	Sergeant	Bronze Medal for Military Valour (Italian)
Matthews, F.	Lieutenant	MC
Mills, E.H.W.	Lieutenant	MC & Bar
Morrison, G.H.	Private	MSM
Moseley, A.E.	Lieutenant	MC
Newport, W.H.	Corporal	MM
Norris, A.R.	Captain	MC
O'Rourke, F.	Private	MM
Owens, L.	Sergeant	MM
Pether, F.	Private	MM
Quinlivan, J.A.	Squadron Sergeant Major	MM
Ranclaud, E.B.	Major	MC
Richardson, E.	Private	MM
Smith, A.R.	Corporal	MM
Spoll, A.A.	Lance Corporal	MM
Steele, C.	Private	MM
Wright, C.	Captain	MC, Serbian Order of the White Eagle, 5th Class (with swords)

In addition 30 Mentions in Despatches were awarded to members of the 1st Battalion 1CB/14th Light Horse Regiment.

(We regret many omissions due to the difficulty of tracing records.)

Appendix 6

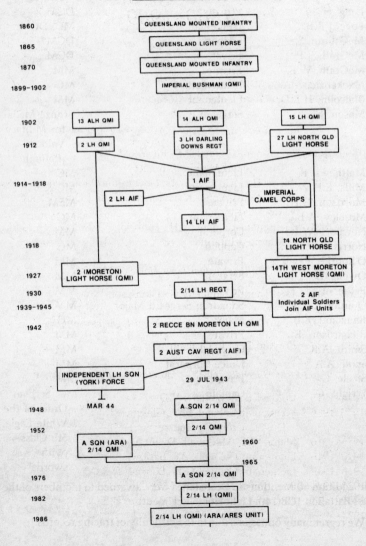

Appendix 7

Commanding Officers 1860-1989

1860-1863 Captain J. Bramston
1884-1899 Lieutenant Colonel P.R. Ricardo

South Africa Contingents 1899-1902

1899-1900 Lieutenant Colonal P.R. Ricardo – 1st Queensland Mounted Infantry
1900-1901 Major H.G. Chauvel – 1st Queensland Mounted Infantry
1900-1901 Lieutenant Colonel K. Hutchison – 2nd Queensland Mounted Infantry
1900-1901 Lieutenant Colonel W.H. Tunbridge – 3rd Queensland Mounted Infantry

15th Australian Light Horse (QMI)

1906-1912 Lieutenant Colonel W.G. Thompson

27th Light Horse – North Queensland Light Horse (QMI)

1912-1917 Captain D.M.L. Graham

2nd Light Horse Regiment – Moreton Light Horse (QMI)

1914-1915 Colonel R.M. Stodart, V.D.
1915-1916 Lieutenant Colonel T.W. Glasgow, K.C.M.G., C.B., D.S.O., Croix De Guerre
1916-1916 Major W.E. Markwell, D.S.O.
1916-1916 Lieutenant Colonel S.W. Barlow
1916-1919 Lieutenant Colonel G.H. Bourne, D.S.O.
1920-1921 Brigadier R.M. Stodart, V.D.
1921-1924 Lieutenant Colonel G.H. Bourne, D.S.O.
1924-1928 Major A. Martin, V.D.

1928-1929   Lieutenant Colonel W.E. Holmes
1929-1930   Major H. Hackney, M.C.

14th Light Horse Regiment — West Moreton Light Horse (QMI)

1916-1918   Lieutenant Colonel G.F. Langley, D.S.O.
1918-1919   Major A.S. Nobbs
1919-1921   Major H.R. Denson, D.S.O.
1921-1923   Lieutenant Colonel Sir Donald Cameron, C.M.G., D.S.O.
1923-1925   Major W.C. Logan
1925-1927   Major W. Patrick
1928-1930   Lieutenant Colonel W. Patrick

2nd/14th Queensland Mounted Infantry

1930-1934   Lieutenant Colonel W. Patrick
1934-1938   Lieutenant Colonel H. Hackney, M.C.
1938-1943   Lieutenant Colonel F.C.E. From
1949-1950   Major F.K. Durbridge
1950-1951   Major D. Glasgow
1951-1955   Lieutenant Colonel G.N. Godsall
1956-1958   Lieutenant Colonel W.Q.A. Nichol
1959-1963   Lieutenant Colonel C.J. Millar, E.D.
1964-1964   Lieutenant Colonel B.G. Ranking
1964-1967   Lieutenant Colonel F.J. Morgan
1967-1970   Lieutenant Colonel D.J. Kelly, E.D.
1970-1973   Lieutenant Colonel C.D.F. Wilson, E.D.
1973-1976   Lieutenant Colonel M.W. Farmer
1976-1976   Major G.F. Stapleton
1976-1978   Major D.R. Simpkins
1978-1981   Major T.E. Childs
1981-1982   Lieutenant Colonel J. Oxenham

2nd/14th Light Horse (QMI)

1982-1983   Lieutenant Colonel J. Oxenham
1983-1986   Lieutenant Colonel T.E. Childs
1986-1986   Lieutenant Colonel P.M. Coleman
1987-1988   Lieutenant Colonel C.E. Stephens
1988-   Lieutenant Colonel D. P. Leslie

Notes to the Text

Queensland Colonial Soldiers

1. *Queenslander*, 4 November 1899.
2. *Moreton Bay Courier*, 9 February 1860.
3. *Votes and Proceedings of Queensland Parliament*, 1891 Vol 2, "Special Service by Corps of Queensland Defence Force", p. 5.
4. Ibid, p. 6.
5. *The Gympie Times*, 5 March 1946 and 17 October 1961.

Soldiers of the Queen

1. Published works of Banjo Paterson.
2. Jan Smuts, 1899.
3. *Queenslander*, 4 November 1899.
4. Conan Doyle, *The Great Boer War*, p. 153.
5. K. Denton, *For Queen and Commonwealth*, p. 78.
6. J.C. Ridpath and E.S. Ellis, *The Story of South Africa*, p. 440.
7. R.L. Wallace, *The Australians at the Boer War*, p. 82.
8. Ibid., p. 84.
9. Ibid., p. 123.
10. Ibid., p. 122.
11. Ibid., p. 141.
12. J.M.H. Abbott, *Tommy Cornstalk*.
13. *Sydney Morning Herald*, 4 April 1900.
14. *Sydney Morning Herald*, 28 May 1900.
15. Captain R.B. Echlin to the Colonial Secretary, 17 August 1901.
16. Ibid.
17. Cable from British High Commission Capetown, 15 August 1900.
18. G.B. Barton and others, *The Story of South Africa*, Vol. 2, p. 458.
19. Denton, op. cit., p. 138.
20. Ibid., p. 138.
21. Ibid., p. 139.
22. Conan Doyle, op. cit., p. 363.
23. Published works of George Essex Evans.

24. Conan Doyle, op. cit., p. 419.
25. Denton, op. cit., p. 142.
26. Wallace, op. cit., p. 302.
27. Ibid., p. 305.
28. Ibid., p. 357.
29. J. Hardie, African War Veterans (an unpublished letter).

The Great War

1. K. Ataturk, 1934.
2. A.J. Hill, *Chauvel of the Light Horse*, p. 43.
3. G.H. Bourne, *The History of the 2nd Light Horse Regiment*, p. 7.
4. H. Chauvel, letter dated 15 December 1914.
5. Bourne, op. cit., p. 13.
6. Hill, op. cit., p. 50.
7. Reverend G. Green's diaries.
8. Ibid.
9. Bourne, op. cit., p. 15.
10. Ibid., p. 17.
11. Green, op. cit.
12. Lieutenant H.J. Tiddy's diaries.
13. I. Jones, *The Australian Light Horse*, p. 30.
14. Tiddy, op. cit.
15. Ibid.
16. I. Jones, *The Australian Light Horse*, p. 31.
17. G. & E. Langley, *Sand, Sweat and Camels*, p. 36.
18. *Queensland Digger*, 1 April 1927.

Sand, Sweat and Horses

1. J.E. Bourne, *Calendar for 1918 and Diary of the War*.
2. H.S. Gullet, *Sinai and Palestine*, p. 121.
3. Ibid., p. 190.
4. A.J. Campbell's diaries.
5. Gullet, op. cit., p. 131.
6. Campbell, op. cit.
7. Ibid.
8. G.H. Bourne, *The History of the 2nd Light Horse Regiment*, p. 34.
9. I. Jones, *The Australian Light Horse*, p. 52.
10. Ibid., p. 54.
11. Gullet, op. cit., p. 192.
12. Jones, op. cit., p. 54.
13. Gullet, op. cit., p. 209.
14. Jones, op. cit., p. 58.
15. Ibid., p. 58.
16. Bourne, op. cit., p. 39.
17. Ibid., p. 39.

18. Gullet, op. cit., p. 243.
19. Ibid.

Their Finest Hour

1. H.S. Gullet, *Sinai and Palestine*, p. 671.
2. G.H. Bourne, *The History of the 2nd Light Horse Regiment*, p. 41.
3. Ibid., p. 42.
4. Gullet, op. cit., p. 315.
5. Ibid., p. 337.
6. Bourne, op. cit., p. 45.
7. Gullet, op. cit., p. 357.
8. Bourne, op. cit., p. 46.
9. Ibid., p. 47.
10. Ibid.
11. I. Jones, *The Australian Light Horse*, p. 98.
12. Bourne, op. cit., p. 54.
13. War Diary of the 2nd Light Horse Regiment.
14. Gullet, op. cit., p. 591.
15. Bourne, op. cit., p. 60.
16. War Diary, op. cit.
17. Jones, op. cit., p. 135.
18. War Diary, op. cit.
19. Gullet, op. cit., p. 671.
20. Jones op. cit., p. 148.
21. E. Mitchell, *Light Horse*, p. 92.
22. G. & E. Langley, *Sand, Sweat and Camels*, p. 154.
23. Gullet, op. cit., p. 710.
24. Jones, op. cit., p. 152.
25. Gullet, op. cit., p. 754.
26. Jones, op. cit., p. 157.
27. Trooper Bluegum, "The Horses Stay Behind".
28. Gullet, op. cit., p. 790.
29. Ibid., p. 791.
30. Letter from Major General P.C. Palin, HQ 75th Division to 14th Light Horse dated 1 July 1919.
31. Bill Gammage, *The Broken Years*.

Training Hard

1. Trooper Scrymgeour, *Echoes of the Australian Light Horse in Egypt and Palestine 1917-18*.
2. *Daily Mail*, 20 March 1928.
3. Interview with C. Head, 1988.

Out of the Saddle

1. P.F. Hockings in J.E. Bourne, *Calendar for 1918 and Diary of the War*.
2. Address by Colonel F.J. Morgan, 2 November 1974.
3. Chief of the General Staff's letter 481/1980, dated 26 May 1980.
4. Address by Major General Cooke, 29 June 1986.
5. *Courier Mail*, 14 July 1988.

References

Books and Journals

Abbott, J.M.H. *Tommy Cornstalk*. London: Longmans, 1902.
Australasian Bandsman, December 1971.

Barton, G.B. *The Story of South Africa*. Sydney: Oceanic Publishing Company 1899–1902.
Bates, I.B. Major. *Queensland Mounted Units 1860–1940*. Brisbane: Victoria Barracks Museum and Historical Society, 1988.
Bean, C.E.W. ed. *Official History of Australia in the War of 1914–18*. Sydney: Angus and Robertson, 1921–43.
Bourne, G.H. Lieutenant Colonel. *The History of the 2nd Light Horse Regiment*. Tamworth: Northern Daily Leader. 1927.
Bourne, J.E. *Calendar for 1918 and Diary of the War*. Brisbane: Outridge Printing Co. Ltd.

Clark, Rex Major. *First Queensland Mounted Infantry Contingent in the South African War*. Canberra: 1971.
Creswicke, L. *South Africa and the Transvaal War*. Edinburgh: T. and E. Jack, 1900.

Deneys, R. *Commando — A Boer Journal of the Boer War*. London: Faber and Faber, 1929.
Denton, Kit. *Australians at War. For Queen and Commonwealth*. Sydney: Time-Life Australia, 1987.
Doyle, Arthur Conan. *The Great Boer War*. London: Thomas Nelson and Sons, 1903.

Echlin, R.B. Captain. *Report to Colonial Secretary*, Brisbane: 1901.

Echoes of the Australian Light Horse in Egypt and Palestine 1917-18. Collected poems by lighthorsemen. Cairo: Nile Mission Press.

Evans, George Essex. *Collected Verse of George Essex Evans.* Sydney: Angus and Robertson, 1928.

Gammage, Bill. *The Broken Years,* Canberra: Australian National University Press, 1974.

Gullett, H.S., *Official History of Australians in the War of 1914-18. Volume 7, The Australian Imperial Force in Sinai and Palestine.* Sydney: Angus and Robertson, 1923.

Hall, R.J. *The Australian Light Horse.* Blackburn, Victoria: W.P. Joynt, 1968.

Hill, A.J. *Chauvel of the Light Horse.* Carlton, Victoria: Melbourne University Press, 1978.

Hopkins, R.N.L. Major-General. *Australian Armour.* Canberra: Australian War Memorial and Government Publishing Service, 1978.

Idriess, I. *The Desert Column.* Sydney: Angus and Robertson, 1933.

Johnson, D.H. *Volunteers at Heart. The Queensland Defence Forces 1860-1901.* Brisbane: University of Queensland Press.

Jones, Ian. *Australians at War, The Australian Light Horse.* Sydney: Time-Life Australia, 1987.

Langley, George and Edmee. *Sand, Sweat and Camels.* Kilmore, 1976.

Macksey, Kenneth. *The History of Land Warfare.* New York: Guinness Superlatives Ltd, 1973.

Mitchell, E. *Light Horse.* Melbourne: The Macmillan Company of Australia Pty Ltd, 1978.

Murray, P.L. Lieutenant Colonel. *Australian Military Contingents to the War in South Africa*. Melbourne: Australian Government Printer, 1911.

New South Wales Military Historical Society, *Supplement to Despatch*, September 1971.

Votes and Proceedings of Queensland Parliament 1891.

RSL Handbook, Anzac Jubilee issue 1965.

Ridpath, J.C. and Ellis, E.S. *The Story of South Africa*. Sydney: Oceanic Publishing Company, 1899.

Royal Historical Society of Queensland, *Special Journal*, Volume VI, Number 1. September 1959.

The Times History of the War in South Africa 1899-1902. London: Sampson, Low, Marston and Company Ltd, 1902.

Wallace, R.L. *The Australians at the Boer War*. Canberra: Australian War Memorial and the Government Publishing Service, 1976.

Newspapers

Courier Mail, 14 July 1988.
Daily Mail, 20 March 1928.
Gympie Times, 1946 and 1961.
Moreton Bay Courier, 1860.
Sydney Morning Herald, correspondence of Banjo Patterson, 1899- 1902.
The *Worker*, March 1890 to June 1891.
Queenslander, 1899.
Queensland Digger, 1 April 1927.

Soldiers' letters published in the *Sydney Morning Herald, Daily Telegraph, Age, Argus, Courier,* and *Advertiser, 1899-*1902.

Unpublished Material

Campbell, A.J. Personal diaries from the First World War.
Cooke, K.J. Major-General. Transcript of address to 2nd/14th Light Horse (Queensland Mounted Infantry). Enoggera: 29 June 1986.
Chief of the General Staff letter No. 481/1980, dated 26 May 1980.

Green, G. Reverend. Personal diaries from the First World War.
Gregory, K. Unpublished letter to authors, 8 November 1985.

Hardie, Jean. Unpublished letter relating to South African War Veterans Association.
Head, C. Interviews with authors, 1986–88.

Morgan, F.J. Colonel. Transcript of address to 2nd/14th Light Horse (Queensland Mounted Infantry). Brisbane: 2 November 1974.

Palin, P.C. Major General. Letter to 14th Light Horse, dated 1 July 1919.

Reading, Bill. Interviews with authors 1986 to 1989.
Record of casualties of 2nd Light Horse Regiment from September 1914 to April 1919.

Tiddy, H.G. Major. Personal diaries from the First World War.

War diary of the 2nd Light Horse AIF from 18 August 1914 to 30 April 1919.

Every effort has been made to contact and acknowledge owners of copyright material used in this book. In case of an omission, holders of copyright are invited to contact:

2nd/14th Light Horse (Queensland Mounted Infantry)
Chauvel Drive
Enoggera Barracks
Qld 4052

Index

Abbott, Corporal J.H.M., 30
Abu Tellul, 132-40, 191
Adie, Lieutenant Alfred, 18, 23, 31
Airey, Colonel, 36
Albany, 57
Alexandria, 57, 72, 95
Alford, Private, 49
Allenby, General Sir Edmund, 111, 113, 118, 120, 121, 123, 129, 131, 140-41, 143-44, 152-53, 155-56
Amman, 124-25, 129, 141-43, 152-53
Annat, Lieutenant James, 41
Anzac Cove. *See* Gallipoli
Anzac Mounted Division, 82, 92, 97, 102, 107, 119, 120, 123-25, 130, 140, 143, 152, 155, 158
Apelt, Corporal, 123
Appleby, Major Syd, 169, 187
Asluj, 114, 116-17
Atkins, Ernest, 186
Australian Commonwealth Horse Regiment, 16
Australian Units –
  Anzac Mounted Division, 82, 102, 107, 111, 118, 120, 124-25; at Romani, 92; advance to Amman, 140-43; achievements, 152
  Australian Mounted Division, 118-21, 130, 145-46, 148, 151-52, 155, 158
  1st Cavalry Brigade, 163, 172
  Light Horse Brigades (AIF) –
    1st Brigade, 68, 82, 95, 116, 119, 123, 142; formation, 59; defence of Romani, 86-88; attack at Magdhaba, 100; at Rafah, 102; defence of Abu Tellul, 133-40
    2nd Brigade, 82, 97, 99, 102, 152; defence of Romani, 91-95
    3rd Brigade, 68, 81-82, 97, 99, 102, 108, 144, 149, 152, 158; defence of Romani, 88-95
    4th Brigade, 130; charge at Beersheba, 117
    5th Brigade, 141; capture of Tul Keram, 146; destruction at Barada Gorge, 148-49
  Infantry Battalions –
    15th Battalion (AIF), 62-65
    33rd Battalion, 164
  Light Horse Regiments (AIF) –
    1st Regiment, 59, 98, 102, 109, 119-20, 158; formation, 55; at Gallipoli, 62-63; at Romani, 91; at Beersheba, 116; defence of Abu Tellul, 132-36
    2nd Regiment, 59-60, 72, 76-78, 85, 97, 119-20, 130, 141-43, 150-53, 163-65, 92; formation, 55-57; at Gallipoli, 62-71; defence of Romani, 88-94; Magdhaba, 100-103; Gaza and Beersheba, 107-16; into the Jordan Valley, 123-24; defence of Ghoraniye bridgehead, 125-28; defence of Abu

Tellul, 132-40; home, 157-58
3rd Regiment, 59, 102, 116; formation, 55; at Romani, 88-91; Jordan Valley, 123-25; defence of Abu Tellul, 132-36
4th Regiment, 107; charge at Beersheba, 117
5th Regiment, 82, 125, 142
7th Regiment, 92, 153
8th Regiment, 68
9th Regiment, 68
10th Regiment, 68, 145, 149
11th Regiment, 77, 107, 163
12th Regiment, 107, 117
14th Regiment, 55, 110, 141, 145, 152, 163-65, 187; formation, 132; destruction at Barada Gorge, 148; Egyptian Rebellion, 158-59
15th Regiment, 132, 145
Queensland Imperial Bushmen
4th Contingent, 16
5th Contingent, 16; battle at Onverwacht, 49-51
6th Contingent, 16
Queensland Mounted Infantry –
1st Contingent, 16, 20, 33-34; at Sunnyside, 21-24; at Paardeberg Drift, 27-29; at Sanna's Post, 30-32
2nd Contingent, 16, 33; at Sanna's Post, 30-32; relief of Mafeking, 34
3rd Contingent, 16; relief of Mafeking, 34; at Koster River, 36-37; battle of Elands River, 37-47
2nd Australian Cavalry Regiment (AIF), 172
2nd Light Horse (QMI), 163-65
2nd (Moreton) Light Horse (QMI), 172
2nd Reconnaissance Battalion Moreton Light Horse (QMI), 172
2nd/4th Armoured Regiment, 172

2nd/7th Divisional Cavalry Regiment, 171-72, 176
2nd/14th Light Horse Regiment, 165, 167, 170-72, 176
2nd/14th Light Horse (QMI), 2, 24, 52, 189-92
2nd/14th Queensland Mounted Infantry, 180-89
3rd/4th Cavalry Regiment, 190
3rd/9th South Australian Mounted Rifles, 183
4th Cavalry Regiment, 185, 188
4th/19th Prince of Wales Light Horse, 183
14th (West Moreton) Light Horse, 163-65
14th Light Horse (Machine Gun) Regiment, 172
14th Motor Regiment, 172
27th (North Queensland) Light Horse, 164
Alwyn, Lieutenant, 110
Aysut, 76-78

Bach, Miss, 37
Baden-Powell, Lieutenant Colonel R.S.S., 35, 36, 42
Bailey, Colonel P.J., 165
Barada Gorge, 148-49
Barcaldine, 5-6, 8
Barlow, Warrant Officer D.B., 183
Barlow, Major S.W., 76
Barrow, General, 148
battle honours, 55, 165
Baxter, Sergeant Norm, 182
Bayly, Major, 24
Beaudesert, 171, 180
Bedouins, 103-4, 150, 155
Beenleigh Troop, 189-90
Beersheba, charge at, 113-17
Belmont, 21, 25-26
Bethlehem, 121, 123, 131
Betts, Alf, 169
Billy the Bastard, 59, 89
Birdwood, General, 65
Birkbeck, Major, 88-91
"Black Week", 18-19, 25
Blacket, Lance-Corporal, 113
Bloemfontein, 27, 29-30, 33
Bluegum, Trooper, 154

Boer, the, 191
    the battle of Paardeberg Drift, 27–29
    the battle at Sunnyside, 23–25
    British defeated by, 18–19
    capitulation of Bloemfontein, 29–30
    changing nature of war, 47–49
    conflict with Britain, 13–14
    Elands River, 37–45
    fall of Pretoria, 32–33
    Koster River, 36–37
    military strength of, 15, 19
    and Onverwacht, 49–51
    origins of, 11
    relief of Kimberley, 26–27
    relief of Mafeking, 34–35
    relief of Sanna's Post, 30–32
    surrender of, 51
    territorial extent of, 13
Boonah Troop, 168–71
Botha, General, 33, 49
Bourne, Lieutenant Colonel G.H., 56, 60, 63, 65, 67, 69, 88, 90–91, 102–3, 108, 113, 128, 134, 136, 137–38, 140, 163–64
Bowen, Sir George, 2
Boyd, Colonel, 25–26
Boyd, Major C.V., 171
Boyd, Lieutenant, 65
Bramston, Captain John, 2
Breydon, Sergeant-Major Richard, 32
Bridges, General Sir William, 59
Brisbane, 1–2, 52, 56, 157, 180, 189
Brisbane Mounted Infantry, 3
Britain, 1, 11, 73, 95, 101, 103–4, 107–11, 118, 121, 129, 140–41, 145–46, 152, 158
    conflict with Boers, 11, 13–14, 18–19
    criticism of, over South Africa, 15, 18–19
British South Africa Company, 14
Broadwood, Brigadier, 30
Brown, Captain W.J., 127–28, 134
Bulawayo, 35
Bulfin, Lieutenant General Sir E.S., 159
Buller, Sir Redvers, 18–19

Bundaberg Mounted Rifles, 3
Burge, Lieutenant, 69
Busby, Bugler William, 51
Butler, Private, 23
Butters, Captain, 43

Camel Corps Training, 73, 74
Camel Transport Corps, 74
camels. *See* Imperial Camel Brigade; Imperial Camel Corps
Cameron, Lieutenant Colonel, 192
Campbell, Captain, 110
Campbell, Sergeant Alan J., 85, 89–91
Cape Colony, 13–15, 51
Capella, 6
Cape Town, 18, 20, 34
Carlyon, Sergeant, 113, 136
Carrington, General, 38, 41, 42
Chambers, Major, 109
Charleville, 6–7
Charters Towers Mounted Infantry, 3, 6
Chauvel, Lieutenant General Harry, 6–7, 16–17, 20, 23–24, 27, 34, 57, 59–61, 62, 69, 82–104, 111, 117, 118, 120, 129–31, 133, 140–41, 144, 147–48, 150–51, 192
Chaytor, Major-General Sir E.W.C., 84, 140, 142–43, 152, 157
Chess Board, 63, 69
Chetwode, Major-General Sir P.W., 97, 99, 101–3
Chunuk Bair, 67, 69
Citizen Military Forces (CMF), 56, 179, 181, 183
Clermont, 6
Colenso, 19
Commonwealth Military Forces, 55
Community Service, Past and Present, 169
Compton, Captain James, 182
concentration camps, 48, 51
Cook, Joseph, 56
Cooke, Major General K.G., 190
Coppin, Lieutenant Bob, 166
Cornwall, 17–18
Coulter, Lieutenant J., 167
Cox, Brigadier, 98, 100, 102, 138
Cox, Lieutenant, 146

Index

Cronje, General, 27–28
Culliford, Private, 49

Daily Mail, 25
Damascus, battle for, 147–52
Darling Downs Mounted Infantry, 6–7
Darvell, Lieutenant Pat, 169
Defence Act (1884), 3
Delagoa, 32–33
De la Rey, General, 26, 38, 43–44, 47–48
Desert Column, 102
Desert Mounted Corps, 114, 141, 151
Devanna, SS, 62
de Wet, General, 26–28, 30, 34, 47, 49
"Diamond Dollar 86" exercise, 190–91
Dobell, Lieutenant-General Sir C.M., 96, 107–8, 111
Dods, Captain Joseph, 32
Dowse, Captain Richard, 23, 31–32
Doyle, Sir Arthur Conan, 19, 45
Drury, Colonel, 9
Drysdale, Sergeant, 94
Duka, Surgeon-General Albert, 39
Duncola, HMAT, 159
Dundee, 14
Dutch East India Company, 11

Earle, Lieutenant Colonel R.M., 188
Easton, Corporal, 94
Echlin, Captain Richard, 36
Egypt, 57, 60–61. *See also* Light Horse Regiment; Turkey
Egyptian Expeditionary Force, 121
Egyptian Rebellion, 158–59
Elands River, 36, 37–45, 191
El Arish, 96–98, 100, 101. *See also* Light Horse Regiment; Turkey
Elliot, Trooper, 63
El Minya, 76
Emir Said, 149
emu plume, 9, 26, 57
Esani, 114
Es Salt, 129–31, 142
Evans, Captain F., 134, 137, 139
Evans, George Essex, 45–47

Farmer, Colonel M.W., 186
Fethers, Veterinary Officer G.E., 154
5th British Yeomanry Brigade, 81–82, 144
1st Australian Light Horse Squadron, 174–76; *See also* Australian Units
Forbes, Bugler Arthur, 37
Franklin, Major R.N., 141, 150, 157
Freedom of the City, 186–87
French, Lieutenant Colonel George Arthur, 3, 8
French, Major General, 26–27
From, Lieutenant Colonel F.C.E., 187

Gaafar Pasha, 143
Gallipoli, 57, 61, 153, 172
  casualty rates at, 62
  Light Horse at, 62–72
Gammage, Bill, 160
Gaza, first and second battles of, 107–10. *See also* Light Horse Regiment; Turkey
Geddes, Sergeant, 112, 130
German Officer's Trench, 63, 69
Germany, 11, 19, 56–57, 60, 110, 117, 121, 134, 136–39, 144–45, 149
Ghoraniye, battle for, 125–29. *See also* Light Horse Regiment; Turkey
Given, Lieutenant, 100
Glascow, Major, 71, 76
Graham, Major, 65
Grant, Brigadier, 117
Great War. *See* Gallipoli; Germany; Light Horse Regiment; Turkey
Green, Reverend George, 62–63, 66
Green, Reverend James, 42
Greenway, Lieutenant Gordon, 169
Gretna Green races, 168
Guidons, 164–65, 185
Gullet, H.S., 83, 94–95, 98, 104, 110, 113, 128, 140, 156
Gympie Mounted Rifles, 3, 9

Hamilton, Sir Ian, 55–56, 61
Hamilton, William, 8
Hardie Jean, 52

Hardy, Justice, 8
Harrisville Light Horse Troop, 166
Hasluck, Sir Paul, 185
Head, Col, 169
Henderson, Lieutenant L.J., 137-39
Herman, Private, 23
Hinton, Lieutenant, 65, 70
History of 2nd Light Horse Regiment Australian Imperial Force August 1914 to April 1919, 56
Hod el Enna, 84, 88
Hogue, Lieutenant, 71
Hogue, Major O., 147, 149
Hore, Lieutenant Colonel, 38, 43-44
horses, 20-21, 26, 29, 39, 44, 51, 59, 109, 118, 153-54, 172, 175-76
"The Horses Stay Behind", 154
Hotchkiss Automatic Rifle, 108-9, 131, 139, 165
Houston, Padre, 132
Hudson, Captain, 98
Hunter, General, 34
Hutchison, Miss, 52
Hutchison, Colonel K., 16
Hutton, Major-General Sir Edward, 55

Imperial Camel Brigade (ICB), 73-75, 97, 102, 124, 131
Imperial Camel Corps (ICC), 72-73, 76, 132
Imperial Mounted Division, 107
Ingold, A.E., 186
Ionian, 73
Ismet, 114, 116-17

Jackson, Major, 6
Jackson, L.T., 186
Jaffa, 121, 141, 153
Jardine-Vidgen, Captain Graham, 187
Jemmameh, 119-120
Jericho, 123
Jifjafa, 81
Jones, Alderman Clem, 186
Jones, Private Victor, 23, 24
Jordan Valley, 121-24

"Kangaroo 89" exercise, 191

Kantara, 78, 81, 88, 92, 95-96
Katia, 81-84, 86, 93-94
Katib Gannit, 84, 88
Kellie, Captain Charles, 35
Kemp, Lieutenant H.C., 127-28
Khalasa, 114, 117
Khuweilfen, 118-19
Kimberley, 14, 19, 24, 26-27, 191
King, Archibald, 186-87
King, Lieutenant, 130, 136-39
King's Banner, 55
Kitchener, Field Marshal, Earl, 28, 44, 71
Knyvitt, Sergeant-Major Frank, 51
Koster River, 35-37
Kruger, President, 14, 33

Ladysmith, 14, 19
Langley, Lieutenant Colonel G.F., 97, 110, 128-29, 145
Lawrence, T.E. (of Arabia), 75, 142, 148, 150
Lawrence, General, 82, 83, 93-94, 96-97
Leatch, Private, 78
Lemmer, General, 35
Letch, Lieutenant, 9
Lieshman, Bill, 103
Light Horse Regiment(s)
    and Amman, 124-25
    as anti-tank regiment, 183-84
    the Armistice, 151-53
    the Band, 182-83
    battle for Ghoraniye, 125-29
    becomes integrated unit, 190-92
    Beenleigh Troop, 189-90
    Boonah Troop, 168-71
    capture of Damascus, 147-51
    charge at Beersheba, 113-18
    and compulsory military service, 163
    creation of, 55, 131-32
    defence of Abu Tellul, 132-40
    destruction of horses, 153-54
    Egyptian campaign, 57-61, 76-78, 158-59
    El Arish, 96-98
    Es Salt raid, 129-31
    first battle of Gaza, 107-8
    and Gallipoli campaign, 62-72
    granted Freedom of City, 186-87

the Great Ride, 143-46
incident at Surafend, 155-56
intra-regimental competition, 166-68
and Jordan Valley campaign, 121-51
at Magdhaba, 98-101
mechanisation of, 172-73, 179-80
and National Service, 181-83
in New Guinea, 172
at Rafah, 101-4
reforms within, 180-81
renaming of units within, 164-65, 172, 183-84
re-siting of, 185
second battle of Gaza, 108-10
Sinai campaign, 81-95
structure of, 60
unlinking of, 188-89
and World War II, 171-76
York Force, 174-76
Light Horse Units. *See* Australian Units
Linan, Sergeant, 151
Logan, Major T., 69-70
Lone Pine plateau, 67-68
Lord Foster Cup, 167-68
Lord Lamington Shield, 166
Love, Les, 171

Ma'adi, 59
Macarthur-Onslow, Brigadier G., 132, 145-46
McConnell, Troop Leader Keith, 170
McDougall, Captain M.D., 134, 139
McGregor-Lowndes, Mrs Jean, 52, 187
Machin, Captain, 109
Mackay Mounted Rifles, 3, 6
McLean, Captain S.N., 127-28
McLeod, Private David, 24
McLeod, Jean, 52
Mafeking, 14, 19
concentration camp at, 48
relief at, 34-35
Magdhaba, 98-101, 158. *See also* Light Horse Regiment; Turkey
Magersfontein, 19
Magnus, Major Dick, 189

Marakeb, 112
Markwell, Major W.E., 100, 116
Masterton, Private John, 39
Mazar, 97-98
McConnell, Troop Leader Keith, 170
McDougall, Captain M.D., 134, 139
McGregor-Lowndes, Mrs Jean, 52, 187
McLean, Captain S.N., 127-28
McLeod, Private David, 24
McLeod, Jean, 52
Melbourne, 56
Mercer, Signaller, 103
Monash Gully, 63, 68
Moreton Bay Courier, 2
Morgan, Colonel F.J., 186-87
Morrish, Lieutenant Jack, 169
Moseley, Lieutenant, 145
Mount Meredith, 88, 90, 92
Mount Tamborine, 170
Mudros, 71, 72
Murray, Sir Archibald, 78, 81-83, 86, 93, 94, 96
Musallabeh, 133-38, 140

Nablus, 144-46
Natal, 13, 15, 51
National Patriotic Fund, 17
National Service, 181-83
Needham-Walker, Warrant Officer D.R., 76-78
Nek, the, 68
New Zealand, 20, 27, 48-49, 57, 61-62, 92-93, 98, 103, 111, 114, 116, 123, 150, 152, 155-56
Nobbs, Major A.S., 159
Norris, Lieutenant, 69
North Australian Observer Unit, 173

Oghratina, 82, 93
Ogilvy, Lieutenant, 65
Olden, Major, 149-50
Olifant's Nek, 34, 38
Orange Free State, 13-14, 51
Orient, 34
Onverwacht, 49-51

Paardeberg Drift, battle of, 27-29
Palin, Major General P.C., 159

Papua New Guinea, 174, 176
Paterson, Banjo, 10, 27
Paterson, Sergeant, 72
Picton, J.T., 187
Pilcher, Colonel T.D., 21, 23
Pinnock, Captain Philip, 17
Pledger, Lieutenant G.T., 138-39
Plummer, General, 48-49, 51
Pope's Hill, 63, 66, 68
Power, Sergeant James, 51
Pretoria, 15, 32-33, 35
Prince of Wales Cup, 166-67

Queensland
  and casualties in Boer War, 51
  defence capability at separation, 1
  government insures soldiers' lives, 16-17
  government response to Boer conflict, 15-17
Queensland Defence Force, 3, 16
Queenslander, 16-17
Queensland Imperial Bushmen, 16
Queensland Light Horse, 2, 171. *See also* Light Horse Regiment
Queensland Mounted Infantry, 16, 186, 191
  and Boer conflict, 20, 21, 28, 33, 34, 38
  establishment of, 2
  reorganisation of, 55, 163
Queensland Shearers' Union, 5. *See also* Shearers' Strike
Quinn's Post, 63, 66, 68, 191

Rafah, 101-4, 152, 156
Reading, Captain Bill, 165, 171, 181
Reece, Private Herbert, 32
Reese, Lieutenant Charles, 51
Reid, Sir George, 57
Rhodesia, 14
Ricardo, Lieutenant Colonel Percy, 6, 17, 18, 23, 30-32, 34
Richardson, B.J., 186
Rietfontein, 25
Righetti, Lieutenant A.S., 91
Roberts, Lord, 19-20, 26, 28-29, 32-33, 42, 47
Robinson, Lieutenant Clive, 167
Rogers, Terry, 9

Romani, 81-88, 93-95, 97, 152, 191
Rose, Private, 23
Royal Australian Armoured Corps (RAAC), 179-80, 183-84
Royston, Brigadier, 92
Rule, Private, 49
Rustenberg, 36-37, 48
Ryder, Lieutenant H., 169

Sandy, 59
Sanna's Post, 30-32
Seccombe, Private Norman, 30
Second Australian Imperial Force (2nd AIF), 171, 174
2nd Reconnaissance Battalion Moreton Light Horse (QMI), 172, 174
Senussi, 73, 75-76, 131
Shanahan, Major, 9, 88-90
Shea, General, 124-25
Shearers' Strike, 4-9
Sinclair, Trooper, 151
Singleton, 185
Sinton, Lieutenant, 138-39
Smith, Brigadier C.L., 73, 131
Smith, General, 83, 92
Sohag, 76-78
"Soldiers of the Queen", 18
Sommerville, Corporal, 94
South Africa. *See* Boer; Britain
South African War Veterans Association (Queensland), 52, 182-83, 187
Southport, 169, 182, 187
Stable Belt, 188
Stapleton, Major Gerry, 187
Steele, Captain, 76
Stodart, Lieutenant Colonel R.M., 57, 67, 69, 71, 157, 163
Stodart, Captain, 88
Stopford, General Sir Frederick, 67
Stormberg, 19
Strong, Private Charles, 32
Suez Canal, 60-61, 73, 78
Sunnyside, battle of, 21-25, 191
Surafend, 155-56
Sydney Morning Herald, 42

Taylor, George, 8
Tel el Saba, 116
Thompson, Private Hiram, 48

Thompson, Captain William, 30-31
Tiddy, Lieutenant H.J., 67, 70-71
Toll, Major Frederick, 50-51
Townsville, 172, 176
Tozer, Lieutenant Vivian, 9
Transvaal, 13-14, 51
Tunbridge, Major W.H., 16, 38-39, 43, 49
Turkey, 57, 73, 78, 81
   at Abu Tellul, 132-40
   aftermath of Gaza, 119-21
   at Amman, 124-25
   battle at Ghoraniye, 125-29 51
   charge at Beersheba, 113-18
   at Damascus, 147-51
   and Egyptian campaign, 60-61
   enters Great War, 57
   at El Arish, 96-98
   Es Salt, 129-31
   first battle at Gaza, 107-8
   at Gallipoli, 62-72
   Jordan Valley campaign, 121-51
   at Magdhaba, 98-101
   at Rafah, 101-4
   at Romani, 86-95
   Sinai campaign, 81-96
   second battle of Gaza, 108-10

Ulimaroa, HMT, 157

Valletin, Major, 50-51
Valley of Death, 147
Vereeniging, 51-52

Vigars, Trooper, 145
von Kress, General, 94, 107-8, 116, 118
von Sanders, General, 140, 144

Wacol, 185
Wallack, Lieutenant Colonel, 42
Ward, Trooper, 94
Weir, Major, 138-39
Wellington Ridge, 88, 92-93
White, Luton, 169
Williams, Gordon, 192
Williams, Lieutenant C., 168
Wilson, Brigadier C.D.F., 24, 184-85
Wilson, Brigadier, L.C., 159, 165
Wilson, Lieutenant Colonel, 85
Winoski, Stan, 170-71
Woodyatt, Lieutenant P.S.R., 91
World War I. *See* Gallipoli; Light Horse Regiment; Turkey
World War II, 171-76
Wright, Lieutenant H.S., 138-39

York Force. *See* 1st Australian Light Horse Squadron
York Peninsula, 174

Zeerust, 36, 42
Zouch, Lieutenant, 43
Zilikat's Nek, 34
Ziza, 142